WHAT EVERY ENGINEER SHOULD KNOW ABOUT

DECISION MAKING UNDER UNCERTAINTY

WHAT EVERY ENGINEER SHOULD KNOW

A Series

Founding Editor

William H. Middendorf

Department of Electrical and Computer Engineering
University of Cincinnati
Cincinnati, Ohio

ADDITIONAL VOLUMES IN PREPARATION

WHAT EVERY ENGINEER SHOULD KNOW ABOUT
DECISION MAKING UNDER UNCERTAINTY

John X. Wang
Certified Six Sigma Master Black Belt
Certified Reliability Engineer
Ann Arbor, Michigan

CRC Press
Taylor & Francis Group
Boca Raton London New York

CRC Press is an imprint of the
Taylor & Francis Group, an **Informa** business

First published 2002 by Marcel Dekker, Inc.

Published 2019 by CRC Press
Taylor & Francis Group
6000 Broken Sound Parkway NW, Suite 300
Boca Raton, FL 33487-2742

First issued in paperback 2019

No claim to original U.S. Government works

ISBN 13: 978-0-367-44700-7 (pbk)
ISBN 13: 978-0-8247-0808-5 (hbk)

Visit the Taylor & Francis Web site at
http://www.taylorandfrancis.com

and the CRC Press Web site at
http://www.crcpress.com

Preface

The Roman philosopher Seneca said "Nothing is certain except the past." This statement seems very true for engineering, which faces today's and tomorrow's challenges for technical product design, development, production, and services. Most engineering activities involve decision making in terms of selecting the concept, configuration, materials, geometry, and conditions of operation. The information and data necessary for decision making are known with different degrees of confidence at different stages of design. For example, at the preliminary or conceptual design stage, very little information is known about the system. However, as we progress towards the final design, more and more data will be known about the system and its behavior. Thus the ability to handle different types of uncertainty in decision making becomes extremely important.

Volume 36 of the *What Every Engineer Should Know...* series dealt primarily with decision making under risk. In risk engineering and management, information may be unavailable, but a probabilistic description of the missing information is available. A technical decision in such a case might be that a manufacturing engineer knows the probability distribution of manufacturing process outputs, and is trying to determine how to set an inspection policy. The design response might be to construct a stochastic program and find a minimum cost solution for a known defect rate.

Decision making under uncertainty, by contrast, involves distributions that are unknown. This situation involves less knowledge than decision making under risk. A situation that involves decision making under uncertainty might be that a communications design engineer knows that transmission quality is a function of the antenna design, the frequency, and the background radiation, but is unsure of what the distribution of background radiation will be in the user environment. In this situation the design response might be to collect field data in the user environment to characterize the radiation, so that antenna design and frequency can be chosen.

Decision making also involves a still more profound lack of knowledge, where the functional form is completely unknown, and often the relevant input and output variables are unknown as well. An example of this more profound uncertainty is that of a design engineer who is considering building airplane wing panels out of composite materials, but is uncertain of the ability of the new materials to withstand shock loads, and indeed which design variables might affect shock loads. The engineering design response to this situation might be to start an R&D project that will vary possible input variables (panel thickness, bond angle, securement method, loading, etc.), and determine which, if any, of these variables has a significant effect on shock resistance.

Uncertainty is an important factor in engineering decisions. This book introduces general techniques for thinking systematically and quantitatively about uncertainty in engineering decision problems. Topics include: spreadsheet simulation models, sensitivity analysis, probabilistic decision analysis models, value of information, forecasting, utility analysis including uncertainty, etc. The use of spreadsheets is emphasized throughout the book.

In engineering many design problems, the component geometry (due to machine limitations and tolerances), material strength (due to variations in manufacturing processes and chemical composition of materials) and loads (due to component wearout, imbalances and uncertain external effects) are to be treated as random variables with known mean and variability characteristics. The resulting design procedure is known as reliability-based design. The reliability-based design is recognized as a more rational procedure compared to the traditional factor of safety-based design methods. Chapter 1 presents an overview of the decision making under uncertainty using classical and contemporary engineering design examples.

In Chapter 2, we develop the first set of spreadsheet simulation models illustrated in a Microsoft® Excel workbook to introduce some basic ideas about simulation models in spreadsheets: the RAND() function as a Uniform random variable on 0 to 1, independence, conditional probability, conditional independence, and the use of simulation tables and data tables in Excel. We see how to build some conditional probabilities into a simulation model, and how then to estimate other conditional probabilities from simulation data.

Chapter 3 reviews basic ideas about continuous random variables using a second set of spreadsheet models. Topics: random variables with Normal probability distributions (NORMINV, NORMSDIST), making a probability density chart from an inverse cumulative function, and Lognormal random variables (EXP, LN, LNORMINV). To illustrate the application of these probability distributions, we work through the spreadsheet analyses of a case study: decision analysis at a bioengineering firm.

In Chapter 4 we begin to study correlation in Excel using covariance and correlation functions. We use a spreadsheet model to simulate Multivariate Normals and linear combinations of random variables. The case study for a transportation network is used to illustrate the spreadsheet simulation models for correlation topics.

Chapter 5 shows how conditional expectations and conditional cumulative distributions can be estimated in a simulation model. Here we also consider the relationship between correlation models and regression models. Statistical dependence and formulaic dependence, the law of expected posteriors, and regression models are presented in this chapter.

In Chapter 6, we analyze decision variables and strategic use of information to optimize engineering decisions. Here we enhance our spreadsheet simulation models with the use of Excel Solver. Also, we introduce risk aversion: utility functions and certainty equivalents for a decision maker with constant risk tolerance.

Scheduling resources so that real-time requirements can be satisfied (and proved to be satisfied) is a key aspect of engineering decision making for project scheduling and resource allocation. Consider a project involving numerous tasks or activities. Each activity requires resources (e.g., people, equipment) and time to complete. The more resources allocated to any activity, the shorter the time that may be needed to complete it. We address project scheduling problems using Critical Path Methods (CPM) or probabilistic Program Evaluation and Review Techniques (PERT) in Chapter 7.

Process control describes numerous methods for monitoring the quality of a production process. Once a process is under control the question arises, "to what extent does the long-term performance of the process comply with engineering requirements or managerial goals?" For example, considering a piston ring production line, how many of the piston rings that we are using fall within the design specification limits? In more general terms, the question is, "how capable is our process (or supplier) in terms of producing items within the specification limits?" The procedures and indices described in Chapter 8 allow us to summarize the process capability in terms of meaningful percentages and indices for engineering decision making.

Chapter 9 presents emerging decision-making paradigms including a balanced scorecard decision-making system. The balanced scorecard is a new decision-making concept that could help managers at all levels monitor results in their key areas. The balanced scorecard decision-making system is fundamentally different from project management in several respects. The balanced scorecard decision-making process, derived from Deming's Total Quality Management, is a continuous cyclical process, which also reflects the nature of engineering decision-making process.

As Soren Aabye Kieregaard (1813-1855), a Danish writer and thinker, said, "Life can only be understood backwards, but it must be lived forwards." Decision making under uncertainty is an inherent part of an engineer's life, since the invention, design, development, manufacture, and service of engineering products require a forward-looking attitude.

The author wishes to thank Professor Michael Panza of Gannon University for his very helpful review insights.

John X. Wang

Contents

1

Engineering: Making Hard Decisions Under Uncertainty

The goal of most engineering analysis is to provide information or the basis for decision making. Engineering decision making could range from simply selecting the size of a column in a structure, to selecting the site for a major dam, to deciding whether a hybrid vehicle is a viable option for transportation. However, uncertainties are invariably present in practically all facets of engineering decision making. From this perspective, the modeling and assessment of uncertainty is unavoidable in any decision made during the planning and design of an engineering system.

1.1 CASE STUDY: GALILEO'S CANTILEVER BEAM

Engineering design, analysis, modeling and testing are often built upon assumptions, which can be traced back to engineering during the Renaissance Age. The first proposition that Galileo set out to establish concerns the nature of the resistance to fracture of the weightless cantilever beam with a cantilever beam with a concentrated load at its end. In the *Dialogues Concerning Two New Sci-*

ences, Galileo states his fundamental assumption about the behavior of the cantilever beam of Figure 1.1 as follows:

It is clear that, if the cylinder breaks, fracture will occur at the point B where the edge of the mortise acts as a fulcrum for the lever BC, to which the force is applied; the thickness of the solid BA is the other arm of the lever along which is located the resistance. This resistance opposes the separation of the part BD, lying outside the wall, from that portion lying inside.

Figure 1.1 Galileo's loaded cantilever.

Here, Galileo sees the cantilever being pulled apart at section AB uniformly across the section. Today's mechanical engineers can easily recognize the following errors in the earliest cantilever beam engineering model:

assuming a uniform tensile stress across the section AB

neglecting shear stress

A cantilever is a beam supported at one end and carrying a load at the other end or distributed along the unsupported portion. The upper half of the thickness of such a beam is subjected to tensile stress, tending to elongate the fibers, the lower half to compressive stress, tending to crush them. Cantilevers are employed extensively in building construction and in machines. In a building, any beam built into a wall and with the free end projecting forms a cantilever. Longer cantilevers are incorporated in buildings when clear space is required below, with the cantilevers carrying a gallery, roof, canopy, runway for an overhead traveling crane, or part of a building above.

In bridge building a cantilever construction is employed for large spans in certain sites, especially for heavy loading; the classic type is the Forth Bridge, Scotland, composed of three cantilevers with two connecting suspended spans. Cantilever cranes are necessary when a considerable area has to be served, as in steel stockyards and shipbuilding berths. In the lighter types a central traveling tower sustains the cantilever girders on either side. The big hammerhead cranes (up to 300-ton capacity) used in working on ships that have proceeded from the yards to fitting-out basins have a fixed tower and revolving pivot reaching down to rotate the cantilever in a circle.

Beams that strengthen a structure are subject to stresses put upon them by the weight of the structure and by external forces such as wind. How does an engineer know that the beams will be able to withstand such stresses? The answer to this question begins with the linear analysis of static deflections of beams. Intuitively, the strength of a beam is proportional to "the amount of force that may be placed upon it before it begins to noticeably bend." The strategy is to mathematically describe the quantities that affect the deformation of a beam, and to relate these quantities through a *differential equation* that describes the bending of a beam. These quantities are discussed below.

Material Properties

The amount by which a material stretches or compresses when subjected to a given force per unit area is measured by the modulus of elasticity. For small loads, there is an approximately linear relationship between the force per area (called *stress*) and the elongation per unit length (called *strain*) that the beam experiences. The slope of this stress-strain relationship is the modulus of elasticity. In intuitive terms, the larger the modulus of elasticity, the more rigid the material.

Load

When a force is applied to a beam, the force is called a *load*, since the force is often the result of stacking or distributing some mass on top of the beam and considering the resulting force due to gravity. The shape of the mass distribution (or, more generally, the shape of the load) is a key factor in determining how the beam will bend.

Cross section

The cross section of a beam is determined by taking an imaginary cut through the beam perpendicular to the beam's bending axis. For example, engineers sometimes use "I-beams" and "T-beams" which have cross sections that look like the letters "I" and "T." The cross section of a beam determines how a beam reacts to a load, and for this module we will always assume that the beam is a so-called *prismatic beam* with a uniform cross section. The important mathematical properties of a cross-section are its *centroid* and *moment of inertia*.

Support

The way in which a beam is supported also affects the way the beam bends. Mathematically, the method by which a beam is supported determines the *boundary conditions* for the differential equation that models the deflection of the beam.

Among the most crucial assumptions in the solution of any engineering problem is the assumption of how any particular mode of failure will occur. As discussed before, a beam is said to be cantilevered when it projects outward, supported only at one end. A cantilever bridge is generally made with three spans, of which the outer spans are both anchored down at the shore and cantilever out over the channel to be crossed. The central span rests on the cantilevered arms extending from the outer spans; it carries vertical loads like a simply supported beam or a truss--that is, by tension forces in the lower chords and compression in the upper chords. The cantilevers carry their loads by tension in the upper chords and compression in the lower ones. Inner towers carry those forces by compression to the foundation, and outer towers carry the forces by tension to the far foundations.

Like suspension bridges, steel cantilever bridges generally carry heavy loads over water, so their construction begins with the sinking of caissons and the erection of towers and anchorage. For steel cantilever bridges, the steel frame is built out from the towers toward the center and the abutments. When a shorter central span is required, it is usually floated out and raised into place. The deck is added last. The cantilever method for erecting prestressed concrete bridges consists of building a concrete cantilever in short segments, prestressing each succeeding segment onto the earlier ones. Each new segment is supported by the previous segment while it is being cast, thus avoiding the need for false work.

In Asia, wooden cantilever bridges were popular. The basic design used piles driven into the riverbed and old boats filled with stones sunk between them to make cofferdam-like foundations. When the highest of the stone-filled boats reached above the low-water level, layers of logs were crisscrossed in such a way that, as they rose in height, they jutted farther out toward the adjacent piers. At the top, the Y-shaped, cantilevering piers were joined by long tree trunks. By crisscrossing the logs, the builders allowed water to pass through the piers, offering less resistance to floods than with a solid design. In this respect, these designs presaged some of the advantages of the early iron bridges. In parts of China many bridges had to stand in the spongy silt of river valleys. As these bridges were subject to an unpredictable assortment of tension and compression, the Chinese created a flexible masonry-arch bridge. Using thin, curved slabs of stone, the bridges yielded to considerable deformation before failure.

Figure 1.2 A stone arch bridge.

Engineering builds upon assumptions, which are vulnerable to uncertainties. Galileo's *Dialogues Concerning Two New Sciences* includes what is considered the first attempt to provide an analytical basis for the design of beams to carry designed loads. Galileo also recognized the responsibility of designers in making things correctly or incorrectly. Because Renaissance engineers did not fully understand the principles upon which they were building bridges, ships, and other constructions, they committed the human errors that were the ultimate causes of many design failures. However, Galileo set out to lay the foundations for a new engineering science. This foundation gives today's engineers the analytical tools to eliminate errors from their conceptions and explorations.

1.2 IMPACT OF ENGINEERING DECISION MAKING

The Hyatt Regency Hotel was built in Kansas City, Missouri in 1978. A state-of-the-art facility, this hotel boasted a 40 story hotel tower and conference facilities. These two components were connected by an open concept atrium. Within this atrium, three suspended walkways connected the hotel and conference facilities on the second, third and fourth levels. Due to their suspension, these walkways were referred to as "floating walkways" or "skyways." The atrium boasted 17 000 square ft (1584 m2) and was 50 ft (15m) high. It seemed unbelievable that such an architectural masterpiece could be the involved in the United States' most devastating structural failure in terms of loss of life and injuries.

It was July 17, 1981 when the guests at the brand new Hyatt Regency Hotel in Kansas City witnessed a catastrophe. Approximately 2000 people were gathered to watch a dance contest in the hotel's state-of-the-art lobby. While the majority of the guests were on the ground level, some were dancing on the floating walkways on the second, third and fourth levels. At about 7:05 pm a loud crack was heard as the second-and fourth-level walkways collapsed onto the ground level. This disaster took the lives of 114 people and left over 200 injured.

The failure of the Hyatt Regency walkway was caused by a combination of a few things. The original construction consisted of beams on the sides of the walkway which were hung from a box beam. Three walkways were to exist, for the second, third and fourth floor levels. In the design, the third floor would be constructed completely independent of the other two floor walkways. The second floor would be held up by hanger rods that would be connected through the fourth floor, to the roof framing. The hanger rods would be threaded the entire way up in order to permit each floor to be held

up by independent nuts. This original design was designed to withstand 90 kN of force for each hanger rod connection. Since the bolt
connection to the wide flange had virtually no moment, it was modeled as a hinge. The fixed end of the walkway was also modeled as a
hinge while the bearing end was modeled as a roller.

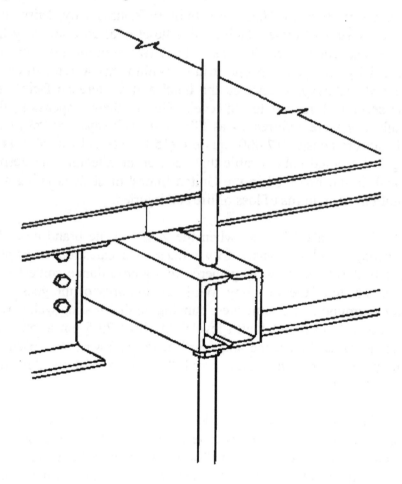

Figure 1.3 The original walkway design.

The new design, created in part to prevent the necessity of requiring the thread to be throughout the entire rod, consisted of one
hanger connection between the roof and the fourth floor and a sec-

ond between the second and the fourth floor. This revised design consisted of the following:

one end of each support rod was attached to the atrium's roof crossbeams;

the bottom end went through the box beam where a washer and nut were threaded on;

the second rod was attached to the box beam 4" from the first rod;

additional rods suspended down to support the second level in a similar manner.

Due to the addition of another rod in the actual design, the load on the nut connecting the fourth floor segment was increased. The original load for each hanger rod was to be 90 kN, but with the design alteration the load was increased to 181 kN for the fourth floor box beam. Since the box beams were longitudinally welded, as proposed in the original design, they could not hold the weight of the two walkways. During the collapse, the box beam split and the support rod pulled through the box beam resulting in the fourth and second level walkways falling to the ground level.

The collapse of the Kansas City Hyatt Regency walkway was a great structural mishap which can be explained in terms of the common results of most structural disasters. In general there exist six main causes for most structural failures.

a lack of consideration for every force acting on particular connections. This is especially prevalent in cases in which a volume change will effect the forces;

abrupt geometry changes which result in high concentrations of stress on particular areas;

a failure to take motion and rotation into account in the design;

"improper preparation of mating surfaces and installation of connections," according to certain engineers who studied the Hyatt Regency case;

a connection resulting in the degrading of materials;

a failure to account for residual stresses arising from manufacturing.

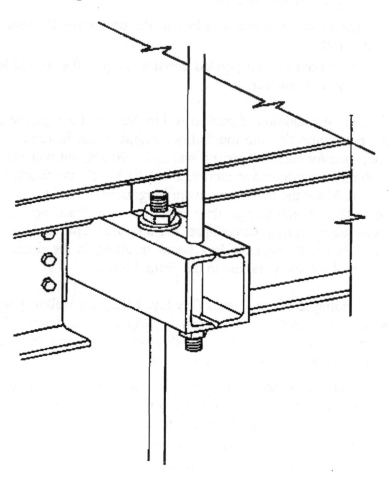

Figure 1.4 The revised walkway design.

Engineering design encompasses a wide range of activities whose goal is to determine all attributes of a product before it is manufactured. A strong capability to engineer industrial and consumer products is needed by any nation to stay competitive in an increasingly global economy. Good engineering design know-how results in lower time to market, better quality, lower cost, lower use of energy and natural resources, and minimization of adverse effects on the environment.

Engineering decision making theory recognizes that the ranking produced by using a criterion has to be consistent with the engineer's objectives and preferences. The theory offers a rich collection of techniques and procedures to reveal preferences and to introduce them into models of decision. It is not concerned with defining objectives, designing the alternatives or assessing the consequences; it usually considers them as given from outside, or previously determined. Given a set of alternatives, a set of consequences, and a correspondence between those sets, decision theory offers conceptually simple procedures for choice.

In a decision situation under certainty the decision maker's preferences are simulated by a single-attribute or *multi-attribute value function* that introduces ordering on the set of consequences and thus also ranks the alternatives. Decision theory for risk conditions is based on the concept of utility. The engineer's preferences for the mutually exclusive consequences of an alternative are described by a utility function that permits calculation of *the expected utility* for each alternative. The alternative with the highest *expected utility* is considered the most preferable. For the case of uncertainty, decision-making theory offers two main approaches. The first exploits criteria of choice developed in a broader context by game theory, as for example the MAX-MIN rule, where we choose the alternative such that the worst possible consequence of the chosen alternative is better than (or equal to) the best possible consequence of any other alternative. The second approach is to reduce the uncertainty case to the case of risk by using *subjective probabilities*, based on expert assessments or on analysis of previous decisions made in similar circumstances.

1.3 UNCERTAINTY AND RISK ENGINEERING

As technology advances, risks are unavoidable. Thus, the issues of risk and decision making confront all engineering professionals. Recognizing there will always be some measure of risk associated with engineering design, how do engineers know when those risks outweigh the possible benefits gained from their work? How do they make informed decisions?

Engineering, more than any other profession, involves social experimentation. Often one engineer's decision affects the safety of countless lives. It is, therefore, important that engineers constantly remember that their first obligation is to ensure the public's safety. This is a difficult assignment, for engineers are not typically autonomous professionals. Most of them work for salaries within a structured environment where budgets, schedules and multiple projects are important factors in the decision-making process.

The decision-making process is often complicated by the fact that most engineers have multiple responsibilities attached to their job descriptions. They are responsible for actual engineering practice (including research, development and design), for making proposals and writing reports, for managing projects and personnel, and often for sales and client liaison. In other words, engineers, by the very nature of their professional stature both outside and inside the corporate structure, cannot work in a vacuum. Graduation is not a license to merely tinker in an engineering laboratory. As an engineer advances, she will be given more authority for directing projects.

This is a natural phenomenon in the engineering community. Most engineers aspire to managerial positions, even if only on one specific project. There are many benefits associated with managerial authority, not the least of which are increased financial remuneration and job satisfaction. But authority comes with a heavy price tag: increased responsibility for decisions made. It is important to remember that responsibility always rests with the project leader.

Eventually, engineers and engineering managers have to make tough decisions about whether a product is safe for public use. Sometimes those decisions involve conflicts over technical problems versus budgets, and problems with schedules and personnel allocations. The engineering manager must first be an engineering professional. Before attending to profits, she must meet professional engineering code requirements and obligations to public safety. This requirement can create difficulties for the engineer.

The problems engineering professionals face involve how to define, assess and manage risk in the light of obligations to the public at large, the employer, and the engineering profession as a whole. The following literature review acts as a catalyst for discussion on risk and the decision-making process as it relates to the cases you are studying. Bear in mind that, above all, risk assessment is closely tied to the perspective that engineering is a social experiment, that engineers have an implicit social contract with the public they serve, and that professional societies and their codes of ethics play important roles in helping shape the engineering decision-making process.

In decision theory and statistics, a precise distinction is made between a situation of risk and one of certainty. There is an uncontrollable random event inherent in both of these situations. The distinction is that in a risky situation the uncontrollable random event comes from a known probability distribution, whereas in an uncertain situation the probability distribution is unknown.

The (average) number of binary decisions a decision maker has to make in order to select one out of a set of mutually exclusive alternatives, a measure of an observer's ignorance or lack of information (see bit). Since the categories within which events are observed are always specified by an observer, the notion of uncertainty emphasizes the cognitive dimension of information processes, specifically in the form of measures of variety, statistical entropy including noise and equivocation.

In decision-making theory and in statistics, risk means uncertainty for which the probability distribution is known. Accordingly, RISK analysis means a study to determine the outcomes of decisions along with their probabilities -- for example, answering the question: "What is the likelihood of achieving a $1,000,000 cost saving in this innovative disk drive?" In systems analysis, an engineer is often concerned with the probability that a project (the chosen alternative) cannot be carried out with the time and money available. This risk of failure may differ from alternative to alternative and should be estimated as part of the analysis.

In another usage, risk means an uncertain and strongly adverse impact, as in "the risks of nuclear power plants to the population are..." In that case, risk analysis or risk assessment is a study composed of two parts, the first dealing with the identification of the strongly adverse impacts, and the second with determination of their respective probabilities. In decision making under uncertainty, risk analysis aims at minimizing the failure to achieve a desired result, particularly when that result is influenced by factors not entirely under the engineer's control.

Risk engineering is an integrated process which includes the following two important parts (Wang and Roush, 2000):

1. Through risk assessment, uncertainties will be modeled and assessed, and their effects on a given decision evaluated systematically;
2. Through design for risk engineering, the risk associated with each decision alternative may be delineated and, if cost-effective, measures taken to control or minimize the corresponding possible consequences.

1.4 DECISION MAKING AND SYSTEM ANALYSIS

Systems analysis is an explicit formal inquiry carried out to help engineers identify a better course of action and make a better decision than she or he might otherwise have made. The characteristic attributes of a problem situation where systems analysis is called upon are complexity of the issue and uncertainty of the outcome of any course of action that might reasonably be taken. Systems analysis usually has some combination of the following: identification and re-identification of objectives, constraints, and alternative courses of action; examination of the probable consequences of the alternatives in terms of costs, benefits, and risks; presentation of the results in a comparative framework so that the engineer can make an informed choice from among the alternatives.

The typical use of systems analysis is to guide decisions on issues such as national or corporate plans and programs, resource use and protection policies, research and development in technology, regional and urban development, educational systems, and other social services. Clearly, the nature of these problems requires an interdisciplinary approach. There are several specific kinds or focuses of systems analysis for which different terms are used: a systems analysis related to public decisions is often referred to as *a policy analysis*. A systems analysis that concentrates on comparison and ranking of alternatives on basis of their known characteristics is referred to as *decision analysis*.

That part or aspect of systems analysis that concentrates on finding out whether an intended course of action violates any constraints is referred to as *feasibility analysis*. A systems analysis in which the alternatives are ranked in terms of effectiveness for fixed cost or in terms of cost for equal effectiveness is referred to as *cost-effectiveness analysis*.

Cost-benefit analysis is a study where for each alternative the time stream of costs and the time stream of benefits (both in monetary units) are discounted to yield their present values. The comparison and ranking are made in terms of net benefits (benefits minus cost) or the ratio of benefits to costs. In *risk-benefit analysis*, cost (in monetary units) is assigned to each risk so as to make possible a comparison of the discounted sum of these costs (and of other costs as well) with the discounted sum of benefits that are predicted to result from the decision. The risks considered are usually events whose probability of occurrence is low, but whose adverse consequences would be important (e.g., events such as an earthquake or explosion of a plant).

The diagnosis formulation, and solution of problems that arise out of the complex forms of interaction in systems, from hardware to corporations, that exist or are conceived to accomplish one or more specific objectives. Systems analysis provides a variety of analytical tools, design methods and evaluative techniques to aid in decision making regarding such systems.

1.5 ENGINEERING DECISION MAKING IN SIX STEPS

Today's engineering design is a team effort. Most engineering decisions fail because of organizational rather than analytical issues. Poor leadership, a faulty problem-solving process, poor teamwork, and lack of commitment are often the underpinnings of failed decision processes. Failed decision processes lead to conflict, loss of credibility, diminished competitive advantage, increased costs, inadequate inclusion of stakeholders, and poor implementation. The six-step process helps resolve conflict and build organizational processes and teams to improve decision-making. Based upon the concepts of interest-based negotiation, decision analysis, breakthrough thinking, and public involvement, the six step process addresses four primary areas: (1) procedural considerations; (2) organizational elements; (3) analytical aspects; and (4) contextual elements. Using decision tools, the six steps enable decision makers to manage expectations, solve problems, avoid classic decision traps, and coach leaders and teams to successful decisions.

The six steps of the process are

1. Ensure leadership and commitment
2. Frame the problem
3. Develop evaluation models and formulate alternatives
4. Collect meaningful, reliable data
5. Evaluate alternatives and make decision
6. Develop an implementation plan

Step 1: Ensure Leadership and Commitment

For a decision process supporting management to succeed, a facilitator should own the process. Lack of leadership support and commitments are primary reasons these management decision processes often fail. Lack of leadership and commitment can occur within the organization or by the facilitator. On an organizational level, lack of commitment is manifested through characteristics such as insufficient allocation of resources to conduct the decision process, hampering open communication, not including the right people at the right time, and not providing true and symbolic support. Poor process leadership and commitment on behalf of the facilitator can also undermine support during the management decision process. With a facilitator, this can occur when a common vision is not created, roles in the decision process are not defined, the decision path is not mapped, or clear, and when measurable expectations are not set. There are other characteristics of ineffective or inefficient leadership and commitment, from both the organizational and facilitator level, but these examples are ones that many individuals have experienced.

The facilitator can ensure the necessary leadership and commitment that is needed for a management decision process to be successful by ensuring that management has:

1. Clearly defined the need
2. Committed sufficient resources for the decision process
3. Set the boundaries for the decision
4. Involved and empowered the right stakeholders
5. Define the roles and responsibilities of the participants

The facilitator needs to provide process leadership and demonstrate commitment by:

1. Helping the group establish a vision
2. Layout the decision process that will be used
3. Establish and sustain teamwork
4. Insure credibility of the decision process

Two decision tools that are instrumental in ensuring leadership and commitment are part of the decision process are development a vision statement and creating a decision map. A vision statement can motivate and inspire a group and creates an image of the desired end product. It focuses the group on the target, where they are ultimately going. It also emphasizes that a solution is attainable and helps the group focus on what it wants to create rather than the problems or challenges that may be present. A decision map creates a road map outlining the path the group will take and breaks the overall decision into a series of smaller, more manageable steps. A decision map or path establishes leadership and alignment by demonstrating knowledge of how the decision process will go forward, thereby eliminating confusion and uncertainty about what will happen next. Credibility and trust in the process and the facilitator are increased because everyone knows what to expect.

Step 2:   Frame the Problem

Another reason decision processes often fail is that the problem the decision process is intended to resolve is poorly or inaccurately defined. In our solution-oriented society, it is all too easy to jump to solutions and not take time to ensure that the problem is accurately and completely defined. By applying a decision hierarchy, the facilitator can help a group accurately define or frame the problem they are assembled to solve. A decision hierarchy or pyramid frames a problem in three ways:

1. At the top, it specifies the known policies, givens, and constraints of the decision, the things that drive and impact the decision but are not changeable.
2. In the middle of the pyramid, it identifies problem areas and uncertainties, the area of focus during the process.
3. At the base of the pyramid, it defines the assumptions and details that are follow-up parts of the decision, areas that are beyond the specific scope of the project, and the details for later determination.

Framing the problem can easily be done using a fishbone diagram and an influence diagram. A fishbone diagram enables the team to focus on the content of the problem, by honing in on the causes of the problem rather than the symptoms of the problem. It also creates a good picture of the collective knowledge of the team about the problem. An influence diagram provides the basis for quantitative decision analysis that can be used to compare alternatives. An influence diagram helps teams identify all factors affecting the decision so that an important influence is not omitted inadvertently. It clarifies relationships between decisions to be made, uncertainties that may unfold after the decision is made, and desired outcomes.

Step 3: Develop Evaluation Models and Formulate Alternatives

An important step in any decision process is to develop models to measure success. Without clear evaluation criteria, it is difficult for a decision process to be applied objectively and for the results of such a decision process to be seen as credible.

Achieving consensus about how success will be measured enables a group to reduce the positional bargaining that typically takes place in collaborative settings and moves them into a deliberation style that is more objective, comprehensive, and defensible. At this stage, alternatives are developed based on the groups vision, framing

of the problem and understanding of the issues requiring consideration.

Two evaluation models that help groups measure success objectively are developing an objectives hierarchy and creating a strategy table. An objectives hierarchy allows a group to graphically
communicate values and illustrate tradeoffs. It also provides an opportunity to compare alternatives and assess a monetary value to the
impacts of decisions. Identifying alternatives that overlap or are not
independent is a main focus. Creating a strategy table enables the
group to display options and consider strategy themes in an organized manner. This approach is excellent when considering multiple
and complex alternatives and ensures that a comprehensive set of
feasible options can be developed.

Step 4: Collect Meaningful, Reliable Data

All management decision processes require information or data. Often however, the information that is collected is not of real use to the
decision making while other information that is critical for effective
decision making is not in the process.
The most telling impact of this is the real cost and resource impacts
of collecting too much, not enough or the wrong information. The
result is a decrease in the credibility of the alternatives developed
and selected. Additionally, there is a need in any decision process to
focus information collection so that only information that is critical
to the decision making process is included. Data overload is a common sentiment of many groups and a typical reason many decision
processes are compromised.

Decision analysis tools can be helpful for identifying what information is meaningful to the process and how it should be collected. Nominal Group Technique (NGT) and swing weighting can
enable groups and the facilitator to effectively determine which information is most important to the decision making process. NGT
allows a group to quickly come to consensus (or identify the lack of
consensus) about the relative importance of issues or problems by

identifying each team member's personal view of the relative importance. This approach allows each team member to rank issues without pressure from dominant or more vocal team members. It also helps the facilitator shift some of the responsibility for the success of the process on the team members by requiring their active, individual participation. Knowing which issues and considerations are most important will enable a group to focus its data collection on those areas of importance.

Swing weighting is another technique that enables a group to evaluate the relative importance of specific information. Defining criteria weights enables a group to express quantitatively the relative value placed on each objective (previously defined during the Objectives Hierarchy) and its performance criteria. Knowing the relative importance of each objective and the issues related to it will enable the group to focus its data collection efforts on those objectives and issues that are most important. Furthermore, by using different sets of weights, a group can represent the views of multiple stakeholders and perform a sensitivity analysis on how alternative viewpoints effect strategy.

Step 5: Evaluate Alternatives and Make a Decision

Once all the alternatives have been created, they must be evaluated and a selection made. Facilitators and groups often have difficulty in this area, if the evaluation effort is not conducted in an organized and logical manner. The credibility of the selected alternative or solution rests on the defensibility of the evaluation process. Two tools that help groups evaluate alternatives and are easy to facilitate are decision matrices and prioritization through cost-benefit analyses.

A decision matrix allows the group to organize its thoughts about each alternative or solution according to criteria defined by the group. Developing and agreeing on the evaluation criteria before discussing how well (or poorly) each alternative meets the criteria is the first step in developing the decision matrix. The evaluation of the alternatives or strategies can now take place in an objective manner. This process allows the facilitator and the group to identify the strengths and weaknesses of proposed alternatives or solutions.

Not all alternatives will achieve project goals in the same way or to the same level. Prioritization through cost-benefit analysis allows the group to compare and contrast alternatives that have different outcomes or achieve different goals. The outcome of this process is a list of items detailing their specific benefits and estimated costs. This method is beneficial when maximization of goals is important.

Step 6: Develop an Implementation Plan

The credibility of any decision making process rests in part on how well the decisions that are made are actually implemented and how effectively the implementation is carried out. An implementation plan moves the team or group from a planning phase to an implementation phase, by linking an action to each goal. An implementation plan also allows the group to consider barriers, performance interventions, and project management issues that could not or were not addressed in the planning phase.

Successful decision process implementation can occur through the use of action plans and decision tree diagrams. An action plan identifies all the needed actions, target deadlines for critical activities, as well as who will complete the action and the estimated cost or resources needed. It provides the group with the ability to layout what should happen and then track what actually happens. Decision tree diagrams allows the group to see the level of complexity associated with implementing an alternative or solution. It moves the team from the planning phase, which often holds implementation considerations at bay, to the real world where implementation issues become the focus. Decision tree diagrams also enable a team to identify needed changes in procedures or at an organizational level.

The six-step process can help a facilitator plan and conduct a collaborative decision making process effectively by identifying decision traps and resolving unproductive group behavior. The process allows facilitators to foster teamwork, use facts to enhance credibility, and manage risk and conflict.

1.6 CUSTOMER-FOCUSED ENGINEERING DECISION-MAKING SYSTEM

Development of a customer focused engineering decision-making system requires an established total quality environment. The focus of the project management system is customer satisfaction.

All project management systems involve:

Analysis

Planning

Implementation

Evaluation

Analysis

The analysis process consists of

1. identifying the target customers
2. determining customer wants, needs, and expectations
3. defining how the organization must adapt to changing customer requirements
4. evaluating customer and supplier relationships
5. determining the processes in the organization that are needed to meet customer expectations
6. assessing management support and commitment
7. assessing the performance of critical processes
8. benchmarking processes
9. judging if process performance is adequate
10. establishing process improvement goals
11. identifying the particular deliverables required for customer satisfaction

12. recognizing risk
13. determining competitive advantage
14. developing metrics
15. performing tradeoffs

Planning

The planning process in a project management system provides tools to help the project team and its resulting deliverable. The planning process provides guidance for

1. relating with the customer
2. preparing proposals
3. planning strategy
4. documenting process information
5. developing mission objectives and goals
6. setting priorities
7. establishing an organizational structure
8. utilizing resources, including people, technology, facilities, tools, equipment, supplies, and funds
9. selecting and training people
10. setting up the project management information system
11. managing the project
12. identifying roles and responsibilities
13. empowering teams and people
14. developing supplier relationships
15. funding the project
16. measuring and reviewing progress
17. designing and developing the deliverable
18. investigating risk
19. solving problems
20. improving processes
21. maintaining accurate configuration information
22. providing and communicating necessary information
23. supporting the deliverable
24. scheduling the work
25. building teamwork

26. closing the project

Implementation

The implementation process equips the project team with approaches to ensure that the project is successful. The implementation process allows the team to

1. set performance measures
2. direct the use of resources
3. handle project changes
4. provide negotiation methods
5. manage risk
6. control costs
7. manage conflict
8. motivate team members
9. take corrective management
10. deliver the outcome (deliverable) to the customer
11. support the deliverable

Evaluation

The evaluation process provides a methodology to assess progress and performance. The evaluation process must provide techniques to

1. measure customer satisfaction
2. document and report the status of cost, schedule, and performance
3. conduct project process reviews
4. keep track of risk
5. test the deliverable
6. gather lessons learned
7. determine the impact on the business
8. continuously improve

1.7   SUMMARY

Engineering decision making theory is a body of knowledge and related analytical techniques of different degrees of formality designed to help an engineer choose among a set of alternatives in light of their possible consequences. Engineering decision-making theory can apply to conditions of certainty, risk, or uncertainty. Decision making under certainty means that each alternative leads to one and only one consequence, and a choice among alternatives is equivalent to a choice among consequences. In decision under risk each alternative will have one of several possible consequences, and the probability of occurrence for each consequence is known. Therefore, each alternative is associated with a probability distribution, and a choice among probability distributions. When the probability distributions are unknown, one speaks about decision making uncertainty.

The objectives of this book are:

to develop in engineers a critical understanding of ideas concerning decision making under risk, uncertainty, ignorance and indeterminacy;

to develop an appreciation that each person and group has knowledge, attitudes and beliefs about risk and uncertainty which, to the individual or group, are 'rational';

to explore the contexts in which engineers manipulate problems involving risk and uncertainty;

to develop a critical appreciation of the uncertainties and subjectivities inherent in modeling;

and to equip engineers with the ability to select and apply appropriate statistical tools, to acquire additional statistical competencies, and to understand their strengths and limitations.

REFERENCES

Ahuja, H. N., Dozzi, S. P., Abourizk, S. M. (1994), Project Management, Second Edition, John Wiley & Sons, Inc., New York, NY.

Ang, A. H-S., Tang, W. H. (1984), Probability Concepts in Engineering Planning and Design, Volume II – Decision, Risk, and Reliability," John Wiley & Sons, New York, NY.

AT&T and the Department of the Navy (1993), Design to Reduce Technical Risk, McGraw-Hill Inc., New York, NY.

Bell, D, E., Schleifer, A., Jr (1995), Decision Making Under Uncertainty, Singular Pub Group, San Diego, California.

Burlando, Tony (1994), "Chaos and Risk Management," Risk Management, Vol. 41 #4, pages 54-61.

Catalani, M. S., Clerico, G. F. (1996), "Decision Making Structures: Dealing With Uncertainty Within Organizations (Contributions to Management Science)," Springer Verlag, Heidelberg, Germany.

Chacko, G. K. (1993), Operations Research/Management Science : Case Studies in Decision Making Under Structured Uncertainty McGraw-Hill, New York, NY.

Chicken, John C. (1994), Managing Risks and Decisions in Major Projects, Chapman & Hall, London, Great Britain.

Cooper, Dale F. (1987), Risk Analysis for Large Projects: Models, Methods, and Cases, Wiley, New York, NY.

Covello, V. T. (1987), Uncertainty in Risk Assessment, Risk Management, and Decision Making (Advances in Risk Analysis, Vol 4), Plenum Press, New York.

Englehart, Joanne P. (1994), "A Historical Look at Risk Management," Risk Management. Vol.41 #3, pages 65-71.

Esenberg, Robert W. (1992), Risk Management in the Public Sector, Risk Management, Vol. 39 #3, pages 72-78.

Galileo (1638), Dialogues Concerning Two New Sciences, Translated by H. Crew and A. de Salvio, 1991, Prometheus Books, Amherst, New York, NY.

Grose, Vernon L. (1987), <u>Managing Risk: Systematic Loss Prevention for Executives</u>, Prentice-Hall, Englewood Cliffs, New Jersey.

Johnson, R. A. (1994), <u>Miller & Freund's Probability & Statistics for Engineers</u>, Fifth Edition, Prentice Hall, New Jersey.

Klapp, M. G. (1992), "Bargaining With Uncertainty: Decision-Making in Public Health, Technologial Safety, and Environmental Quality," March, Auburn House Pub., Westport, Connecticut.

Kurland, Orim M. (1993), "The New Frontier of Aerospace Risks." <u>Risk Management</u>. Vol. 40 #1, pages 33-39.

Lewis, H.W. (1990), <u>Technological Risk</u>, Norton, New York, NY.

McKim, Robert A. (1992), "Risk Management: Back to Basics." <u>Cost Engineering</u>. Vol. 34 #12, pages 7-12.

Moore, Robert H. (1992), "Ethics and Risk Management." <u>Risk Management</u>. 39 #3, pages 85-92.

Moss, Vicki. (1992), "Aviation & Risk Management." <u>Risk Management</u>. Vol. 39 #7, pages 10-18.

Petroski, Henry (1994), <u>Design Paradigms: Case Histories of Error & Judgment in Engineering</u>, Cambridge University Press, New York, NY.

Raftery, John (1993), <u>Risk Analysis in Project Management</u>, Routledge, Chapman and Hall, London, Great Britain.

Schimrock, H. (1991), "Risk Management at ESA." <u>ESA Bulletin</u>. #67, pages 95-98.

Sells, Bill. (1994), "What Asbestos Taught Me About Managing Risk." <u>Harvard Business Review</u>. 72 #2, pages 76-90.

Shaw, Thomas E. (1990), "An Overview of Risk Management Techniques, Methods and Application." <u>AIAA Space Programs and Technology Conference</u>, Sept. 25-27.

Smith, A. (1992), "The Risk Reduction Plan: A Positive Approach to Risk Management." IEEE <u>Colloquium on Risk Analysis Methods and Tools</u>.

Sprent, Peter (1988), <u>Taking Risks: the Science of Uncertainty</u>, Penguin, New York, NY.

Toft, Brian (1994), Learning From Disasters, Butterworth-Heinemann.

Wang, J. X. and Roush, M. L. (2000), What Every Engineer Should Know About Risk Engineering and Management, Marcel Dekker Inc., New York, NY.

Wideman, R. Max., ed. (1992), Project and Program Risk Management: A Guide to Managing Project Risks and Opportunities, Project Management Institute, Drexel Hill, PA.

2

Engineering Judgment for Discrete Uncertain Variables

Engineering judgment is the first and most indis-
pensable tool for engineers. Engineering judgment
not only gets projects started in the right direction,
but also keeps a critical eye on their development,
production, and service. With the helps of computer
simulation, engineering judgment separates the sig-
nificant from the insignificant details for engineer-
ing problems involving uncertainties. Using spread-
sheet on her/his desktop or laptop, every engineer
can do effective simulations to make accurate
judgments.

2.1 CASE STUDY: PRODUCTION OF ELECTRONIC MODULES

The Erie Electronics company produces sophisticated electronic modules in production runs of several thousand at a time. It has been found that the fraction of defective modules in can be very different in different production runs. These differences are caused by micro-irregularities that sometimes occur in the electrical current. For a simple first model, we may assume first that there are just two possible values of the defective rate.

In about 70% of the production runs, the electric current is regular, in which case every module that is produced has an independent 10% chance of being defective.

In the other 30% of production runs, when current is irregular, every module that is produced has an independent 50% chance of being defective.

Testing these modules is quite expensive, so it is valuable to make inferences about the overall rate of defective output based on a small sample of tested modules from each production run. Because of this, 20 modules are tested from a production run. What is the conditional probability that the current is regular given that 5 defective modules are found among the 20 tested?

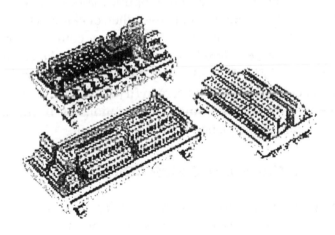

Figure 2.1 Electronic modules - testing is very expensive!

Let's start from the spreadsheet software on our desktop or laptop computers. We will make a simulation model using an Excel spreadsheet to study the conditional probability of irregular current given the results of testing 20 modules from a production run.

Probability is the basic mathematics of uncertainty. Whenever there is something that we do not know for sure, our uncertainty can (in principle) be described by probabilities. This chapter is an elementary introduction to the basic ideas of uncertainty. We will show how to use the ideas of probability to build simulation models of uncertainties in spreadsheets.

Everything that we simulate in spreadsheets will be described with Microsoft Excel 5.0 or higher (in Windows). Within Excel, there are many different ways of telling the program to simulate any one task. Most commands can begin by selecting one of the options on the menu bar at the top of the screen ("File Edit View Insert Format Tools Data Window Help"), and you can make this selection either by clicking with the mouse or by pressing the [Alt] key and the underlined letter key. Then, secondary options appear under the menu bar, and you select among them by either clicking with the mouse or by typing the underlined letter key. Many common command sequences can also be entered by a short-cut keystroke (which is indicated in the pop-down menus), or by clicking on a button in a power bar that you can display on the screen (try the View menu). In this chapter, we will describe command descriptions as if you always use the full command sequence from the top menu.

2.2 HOW WOULD ENGINEERS FLIP COINS?

When we study probability theory, we are studying uncertainty. To study uncertainty with a spreadsheet-based simulation, it is useful to create some uncertainty within the spreadsheet itself. Knowing this, the designers of Excel gave us one simple but versatile way to create such uncertainty: the RAND() function.

Let's see now how to use this function to create a spreadsheet that simulates the following situation: "When current is irregular, every module that is produced has an independent 50% chance of being defective." Since every electronic module is now equally likely to be "good" and "defective," this simulation would behave like a coin toss.

With the cursor on cell A1 in the spreadsheet, let us type the formula

=RAND ( )

and then press the Enter key. A number between 0 and 1 is displayed in cell A1. Then (by mouse or arrow keys) let us move the cursor to cell B1 and enter the formula

=IF(A1<0.5,"Defective","Good")

(The initial equals sign [=] alerts Excel to the fact that what follows is to be interpreted as a mathematical formula, not as a text label. The value of an IF(•,•,•) function is its second parameter if the first parameter is a true statement, but its value is the third parameter if the first is false.) Now Excel checks the numerical value of cell A1, and if A1 is less than 0.5 then Excel displays the text "Defective" in cell B1, and if A1 is greater than or equal to 0.5 then Excel displays the text "Good" in cell B1.

If you observed this construction carefully, you would have noticed that the number in cell A1 changed when the formula was entered into cell B1. In fact, every time we enter anything into spreadsheet, Excel recalculates the everything in the spreadsheet and it picks a new value for our RAND() function. (We are assuming here that Excel's calculation option is set to "Automatic" on your computer. If not, this setting can be changed under Excel's Tools>Options menu.) We can also force such recalculation of the spreadsheet by pushing the "Recalc" button, which is the [F9] key in Excel. If you have set up this spreadsheet as we described above, try pressing [F9] a few times, and watch how the number in cell A1 and the text in cell B1 change each time.

Now let us take hands away from the keyboard and ask the question: What will be the next value to appear in cell A1 when the

[F9] key is pressed next time? The answer is that we do not know. The way that the Excel program determines the value of the RAND() function each time is, and should remain, a mystery to us. The only thing that we need to know about these RAND() calculations is that the value, rounded to any number of decimal digits, is equally likely to be any number between 0 and 1. That is, the first digit of the decimal expansion of this number is equally likely to be 0, 1, 2, 3, 4, 5, 6, 7, 8, or 9. Similarly, regardless of the first digit, the second decimal place is equally likely to be any of these digits from 0 to 9, and so on. Thus, the value of RAND() is just as likely to be between 0 and 0.1 as it is to be between 0.3 and 0.4. More generally, for any number v, w, x, and y that are between 0 and 1, if v - w = x - y then the value of the RAND() expression is as likely to be between w and v as it is to be between y and x. This information can be summarized by saying that, from our point of view, RAND() is drawn from a uniform probability distribution over the interval from 0 to 1.

The cell B1 displays "Defective" if the value of A1 is between 0 and 0.5, whereas it displays "Good" if A1 is between 0.5 and 1. Because these two intervals have the same length (0.5-0 = 1-0.5), these two events are equally likely. That is, based on our current information, we should think that, after we next press [F9], the next value of cell B1 is equally likely to be "Defective" or "Good". So we have created a spreadsheet cell that behaves just like a fair coin toss every time we press [F9]. Now we have our first simulation model.

We can press [F9] a few more times to verify that, although it is impossible to predict whether "Defective" or "Good" will occur next in cell B1, they tend to happen about equally often when we recalculate many times. It would be easier to appreciate this fact if we could see many of these simulated test results at once. This is easy to do by using the spreadsheet's Edit>Copy and Edit>Paste commands to make copies of our formulas in the cells A1 and B1. So let us make copies of this range A1:B1 in all of the first 20 rows of the spreadsheet. (Any range in a spreadsheet can be denoted by listing its top-left and bottom right cells, separated by a colon.)

To copy in Excel, we must first select the range that we want to copy. This can be done by moving the cursor to cell A1 and then holding down the shift key while we move the cursor to B1 with the right arrow key. (Pressing an arrow key while holding down the shift key selects a rectangular range that has one corner at the cell where the cursor was when the shift key was first depressed, and has its opposite corner at the current cell. The selected range will be highlighted in the spreadsheet.) Then, having selected the range A1:B1 in the spreadsheet, open the Edit menu and choose Copy. Faint dots around the A1:B1 range indicate that this range has been copied to Excel's "clipboard" and is available to be pasted elsewhere. Next, select the range A1:A20 in the spreadsheet, and then open the Edit menu again and choose Paste. Now the spreadsheet should look something as shown in Figure 2.2. The descriptions that appear in the lower right corner of Figure 2.2 are description text typed into cells E17:E20.

In Figure 2.2, we have made twenty copies of the horizontal range A1:B1, putting the left-hand side of each copy in one of the cells in the vertical range A1:A20. So each of these twenty A-cells contains the RAND() function, but the values that are displayed in cells A1:A20 are different. The value of each RAND() is calculated independently of all the other RANDs in the spreadsheet. The spreadsheet even calculates different RANDs within one cell independently, and so a cell containing the formula =RAND() RAND() could take any value between -1 and +1.

The word "independently" is being used here in a specific technical sense that is very important in probability theory. When we say that a collection of unknown quantities are independent of each other, we mean that learning the values of some of these quantities would not change our beliefs about the other unknown quantities in this collection. So when we say that the RAND() in cell A20 is independent of the other RANDs in cells A1:A19, we mean that knowing the values of cells A1:A19 tells us nothing at all about the value of cell A20. If you covered up cell A20 but studied the values of cells A1:A19 very carefully, you should still think that the value of cell A20 is drawn from a uniform distribution over the interval from 0 to 1 (and so, for example, is equally likely to be above or below 0.5), just as you would have thought before looking at any of these cell values.

	A	B	C	D	E	F	G	H
1	0.625806	Defective						
2	0.492776	Good						
3	0.763691	Defective						
4	0.422546	Good						
5	0.039134	Good						
6	0.551165	Defective						
7	0.206061	Good						
8	0.93392	Defective						
9	0.239356	Good						
10	0.928207	Defective						
11	0.129362	Good						
12	0.676715	Defective						
13	0.228468	Good						
14	0.998251	Defective						
15	0.037361	Good						
16	0.128982	Good						
17	0.940594	Defective			FORMULAS FROM RANGE A1:D20			
18	0.789908	Defective			A1. =RAND()			
19	0.894309	Defective			B1. =IF(A1<0.5,"Defective","Good")			
20	0.217637	Good			A1:B1 copied to A2:A20			

Figure 2.2 Simulating sample testing of 20 electronic modules when the electronic current is irregular.

Each of the twenty cells in B1:B20 contains a copy of the IF... function that we originally entered into cell B1. If you run the cursor through the B cells, however, you should notice that the reference to cell A1 in cell B1 was adjusted when it was copied. For example, B20 contains the formula

=IF(A20<.5,"Defective","Good")

Excel's Copy command treats references to other cells as relative, unless we preface them with dollar signs ($) to make them absolute. So each of the copied IF functions looks to the cell to the left for the number that it compares to 0.5, to determine whether "Defective" or "Good" is displayed. Thus we have set up a spreadsheet in which cells B1:B20 simulate twenty independent events, each with two equally likely outcomes, just like the flip of a coin!

Now let us change this spreadsheet so that it can do something that you could not do so easily with coins.

Let's simulate testing of 20 electronic modules for a production run when the electric current is regular. In this case every module that is produced has an independent 10% chance of being defective. Into cell B1, enter the formula

=IF(A1<$D$1,"Defective","Good")

(You can use the edit key [F?] to get into the old formula and revise it, simply changing the 0.5 to $D$1.) Next, copy cell B1 and paste it to B2:B20. The dollar signs in the formula tell Excel to treat the reference to D1 as an absolute, not to be adjusted when copied, and so the formula in cell B20 (for example) should now be =IF(A20<$D$1,"Defective","Good"). Now enter any number between 0 and 1 into cell D1. If you enter 0.25 into cell D1, for example, then your spreadsheet may look something like Figure 2.3.

In this spreadsheet, each cell in B1:B20 will display "Good" if the random number to the left is between 0 and 0.9; but it is much less likely that the number will be below, and so we should expect substantially less "Defective" modules than "Good" modules. Using the language of probability theory, we may say that, in each cell in the range B1:B20 in this spreadsheet, the probability of getting a "Defective" module after the next recalculation is 0.1. If we entered any other number p into cell D1, then this probability of getting a

"Defective" module in each B-cell would change to this new probability p, independently of the other B-cells.

More generally, when we say that the probability of some event "A" is some number q between 0 and 1, we mean that, given our current information, we think that this event "A" is just as likely to occur as the event that any single RAND() in a spreadsheet will take a value less than q after the next recalculation. That is, we would be indifferent between a lottery ticket that would pay us $100 if the event "A" occurs and a lottery ticket that would pay us $100 if the RAND()'s next value is less than the number q. When this is true, we may write the equation $P(A) = q$.

	A	B	C	D	E	F	G	H
1	0.762732	Good		0.1				
2	0.513582	Good						
3	0.098928	Defective						
4	0.785364	Good						
5	0.267312	Good						
6	0.483421	Good						
7	0.89123	Good						
8	0.979217	Good						
9	0.098169	Defective						
10	0.327113	Good						
11	0.789852	Good						
12	0.349451	Good						
13	0.168069	Good						
14	0.819238	Good						
15	0.951035	Good						
16	0.327866	Good						
17	0.182912	Good			FORMULAS FROM RANGE A1:D20			
18	0.590926	Good			A1. =RAND()			
19	0.695366	Good			B1. =IF(A1<$D$1,"Defective","Good")			
20	0.400321	Good			A1:B1 copied to A2:A20			

Figure 2.3 Simulating sample testing of 20 electronic modules when the electronic current is regular.

2.3 SYNTHESIZE KNOWLEDGE INTO A SIMULATION MODEL

Incomplete information results in uncertainties for engineering judgment and decision making. To minimize the uncertainty and thus improve the accuracy of judgments and decisions, engineers need to combine information from current observations and prior knowledge from historic data. Accurate assessment is the basis of effective engineering decisions.

In the electronic module production problem, we need to make judgment about whether the electric current is regular and make engineering decisions accordingly. The following information reflects the prior knowledge from historic data:

In about 70% of the production runs, the electric current is regular, in which case every module that is produced has an independent 10% chance of being defective.

In the other 30% of production runs, when current is irregular, every module that is produced has an independent 50% chance of being defective.

To synthesize knowledge into our simulation model, let us begin by inserting a new row at the top of the spreadsheet, into which we write labels that describe the interpretation of the cells below. We can insert a new row 1 by selecting cell A1 and entering the command sequence Insert>Rows. Notice that formula references to cell $D$1 automatically change to cell $D$2 when we insert a new row 1 at the top.

Let us enter the label "tests:" in the new empty cell B1. Below, in the 20 cells of range B2:B21, we will simulate the outcomes of the tests to the 20 sample electronic modules, with a 1 denoting a defective module from this test and 0 denoting a good module from this test. So let us re-edit cell B2, entering the formula

=IF(A1<$D$2,1,0)

into cell B2. Recall that we also have =RAND() in cell A2. To model the other 19 calls, let us copy A2:B2 and paste it to A3:A21. If we enter the number 0.5 into cell D2 then, with our new interpretation, this spreadsheet simulates a situation in which the test has a probability 1/2 of finding a defective module with each of the 20 sample tests. So we should enter the label "P(Defective)" in each call into cell D1. To indicate that the values in column A are just random numbers that we use in building our simulation model, we may enter the label "(rands)" in cell A1.

Indicating defective modules and good modules in the various calls by 1s and 0s makes it easy for us to count the total number of defective modules in the spreadsheet. We can simply enter the formula =SUM(B2:B21) into cell D10 (say), and enter the label "Total Defective Modules" into the cell above. The result should look similar to Figure 2.4.

The Figure 2.4 simulates when electronic current is irregular, every module that is produced has an independent 50% chance of being defective. To simulate 20 sample test when electronic current is regular, we could simply replace the 0.50 in cell D2 by 0.1, representing that every module that is produced has an independent 50% chance of being defective.

However, once the number 0.50 is entered into cell D2 in Figure 2.4, the outcomes of the 20 simulated tests are determined independently, because each depends on a different RAND variable in the spreadsheet. In this spreadsheet, if we knew that we were getting good electric modules from all of the first 19 tests, we would still think that the probability of getting a good electronic module from the 20th customer is 1/2 (as likely as a RAND() being less than 0.50). Such a strong independence assumption may seem very unrealistic. In real life, a string of 19 successful tests might cause us to infer that the electric current for this particular production run is regular, and so we might think that we would be much more likely to get a good module on our 20th test. On the other hand, if we learned that the sample test had a string of 19 unsuccessful tests, then we might infer that the electric current for this particular production run is irregular, and so we might think that it is unlikely to get a good module on the 20th test. To take account of such dependencies, we need to revise our model to one in which the outcomes of the 20 sample tests are not completely independent.

	A	B	C	D	E	F	G	H
1	(rands)	Tests:		P(Defective module)				
2	0.120192	1		0.5				
3	0.621353	0						
4	0.521652	0						
5	0.359255	1						
6	0.455288	1						
7	0.59662	0						
8	0.611503	0						
9	0.778482	0		Total Defective Modules				
10	0.68464	0		8				
11	0.057172	1						
12	0.695656	0						
13	0.246713	1						
14	0.176221	1						
15	0.880846	0						
16	0.995617	0						
17	0.169346	1			FORMULAS FROM RANGE A1:D21			
18	0.36275	1			A2. =RAND()			
19	0.776979	0			B2. =IF(A2<$D$2,1,0)			
20	0.924902	0			A2:B2 copied to A3:A21			
21	0.675541	0			D10. =SUM(B2:B21)			

Figure 2.4 Simulating total defective modules during sample testing when the electronic current is irregular.

This independence problem is important to consider whenever we make simulation models in spreadsheets. The RAND function makes it relatively easy for us to make many random variables and events that are all independent of each other. Making random variables that are not completely independent of each other is more difficult. In Excel, if we want to make two cells not be independent of each other, then there must be at least one cell with a RAND function in it that directly or indirectly influences both of these cells. (We can trace all the cells that directly and indirectly influence any cell by repeatedly using the Tools>Auditing>TracePrecedents command sequence until it adds no more arrows. Then use Tools>Auditing>RemoveAllArrows.) So to avoid assuming that the sample tests simulated in B2:B21 are completely independent, we should think about some unknown quantities or factors that might influence all these tests events, and we should revise our spreadsheet to take account of our uncertainty about such factors.

Notice that the our concern about assuming independence of the 20 sample tests was really motivated in the previous discussion by our uncertainty about whether the electric current is regular. This observation suggests that we should revise the model so that it includes some explicit representation of our uncertainty about the electric current. The way to represent our uncertainty about the electric current is to make the electric current in cell D2 into a random variable. When D2 is random, then the spreadsheet will indeed have a random factor that influences all the 20 sample test events.

In the simulation model for sample testing of production runs, we have two possible levels for electric currents: in about 70% of the production runs, the electric current is regular, in which case every module that is produced has an independent 10% chance of being defective; in the other 30% of production runs, when current is irregular, every module that is produced has an independent 50% chance of being defective.

To model this situation, we can modify the simple model in Figure 2.4 by entering the formula

$$=IF(RAND()<0.7,0.1,0.5)$$

into cell D2. Then we can enter the label "Electric Current" into cell D1, and the result should be similar to Figure 2.5.

If we repeatedly press Recalc [F9] for the spreadsheet in Figure 2.5, we can see the value of cell D2 changing between 0.1 and 0.5. When the value is 0.1, the spreadsheet model is simulating a situation when electric current is regular. When the value is 0.5, the spreadsheet model is simulating a situation when electric current is irregular. When the electronic current is regular, the electronic modules usually fails in far less than half a dozen times; and when it is irregular, the electronic modules usually fails in around half of the tests. But if you recalculate this spreadsheet many times, you will occasionally see it simulating a sample testing when the electronic current is regular but which nevertheless fails in more than half of a dozen times.

100

	A	B	C	D	E	F	G	H
1	(rands)	Tests:		P(Defective module)				
2	0.615774	0		0.1				
3	0.082468	1						
4	0.278071	0						
5	0.146856	0						
6	0.400471	0						
7	0.489316	0						
8	0.785882	0						
9	0.362203	0		Total Defective Modules				
10	0.40328	0		2				
11	0.822602	0						
12	0.982264	0						
13	0.028116	1						
14	0.336278	0						
15	0.793098	0						
16	0.224843	0			FORMULAS FROM RANGE A1:D21			
17	0.395035	0			A2. =RAND()			
18	0.853582	0			B2. =IF(A2<$D$2,1,0)			
19	0.243988	0			A2:B2 copied to A3:A21			
20	0.422008	0			D2. =IF(RAND()<0.7,0.1,0.5)			
21	0.756074	0			D10. =SUM(B2:B21)			

Figure 2.5 Spreadsheet simulating 20 sample testing of electronic modules.

Let us now answer the question that might arise in the work of a supervisor of the production line. If the sample test showed to exactly 5 of the 20 tests for the electronic modules, then what should we think is the probability that the electric current is actually regular (but just had a bad luck for this sample test)? Before starting the sample test we believed that the electric current is regular in about 70% of the production runs; but observing 5 defective modules in 20 gives us some information that should change our assessment about the likelihood for regular current.

To answer this question with our simulation model, we should recalculate the simulation many times. Then we can see how often sample testing of 20 electronic modules gets exactly 5 defective modules when the electric current is regular, and how often sample testing of 20 electronic modules gets exactly 5 defective modules when the electric current is irregular. The relative frequencies of these two events in many recalculated simulations will give us a way to estimate how much can be inferred from the evidence of exactly 5 defective modules out of 20 sample tests.

There is a problem, however. Our model is two-dimensional (spanning many rows and several columns), and it is not so easy to make hundreds of copies of it. Furthermore, even if we did put the whole model into one row of the spreadsheet, recalculating and storing hundreds of copies of 42 different numerical cells could strain the speed and memory of a computer.

But we do not really need a spreadsheet to hold hundreds of copies of our model. In our simulation, we only care about two things: 1) Is the electric current regular; and 2) how many defective modules are found during the sample testing? So if we ask Excel to recalculate our model many times, then we only need it to make a table that records the answers to these two questions for each recalculated simulation. All the other information (about which modules among the twenty in B2:B21 actually failed during the sample testing, and about the specific random values shown in A2:A21) that is generated in the repeated recalculations of the model can be erased as the next recalculation is done. Excel has the capability to make just such a table, which is called a "data table" in the language of spreadsheets.

To make a data table for simulation, the output that we want to tabulate from our model must be listed together in a single row. This model-output row must also have at least one blank cell to its left, and beneath the model-output row there must be many rows of blank cells where the simulation data will be written. In our case, the simulation output that we want is in cells D2 and D10, but we can easily repeat the information from cells in an appropriate model-output row. So to keep track of whether the simulated electric current is regular, let us enter the formula =IF(D2=0.5,1,0) into cell B35, and to keep track of the number of defective modules achieved by this simulated test, let us enter the formula =D10 into cell C35. To remind ourselves of the interpretations of these cells, let us enter the labels 'Current irregular? and '# of Defective Modules into cells B34 and C34 respectively.

Now we select the range in which the simulation data table will be generated. The top row of the selected range must include the model-output range B35:C35 and one additional unused cell to the left (cell A35). So we begin by selecting the cell A35 as the top-left cell of our simulation data table. With the [Shift] key held down, we can then press the right-arrow key twice, to select the range A35:C35. Now, when we extend the selected range downwards, any lower row that we include in the selection will be filled with simulation data. If we want to record the results of about 1000 simulations, then we should include in our simulation table about 1000 rows below A35:C35. So continuing to hold down the [Shift] key, we can tap the [PgDn] key to expand our selection downwards. Notice that the number of rows and columns in the range size is indicated in the formula bar at the top left of the screen while we are using the [PgDn] or arrow keys with the [Shift] key held down. Let us expand the selection until the range A35:C1036 is selected, which gives us 1002 rows that in our selected range: one row at the top for model output and 1001 rows underneath it in which to store data.

After selecting the range A35:C1036, let us use the command sequence Data>Table. When the dialogue box asks us to specify input cells, we should leave the row-input box blank, and we should specify a column-input cell that has no effect on any of our calculations (cell A34 or cell A35 will do). The result will be that Excel then recalculates the model 1001 times and stores the results in the rows of B36:C1036. The column-input substitutions affect nothing, but the recalculation of the RANDs gives us different results in each row. Unfortunately, Excel data tables are alive, in the sense that they will be recalculated every time we recalculate the spreadsheet. To do statistical analysis, we do not want our simulation data to keep changing. To fix this, we should select the data range B36:C1036, copy it to the clipboard (by Edit>Copy), and then with B36:C1036 still selected we should use the command sequence

Edit>PasteSpecial>Values

The result is that the TABLE formulas in the data range are replaced by the values that were displayed, and these numerical values now will not change when [F9] is pressed. The spreadsheet for simulation is shown in Figure 2.6.

Now, recall our question about what we can infer about the electric current if sample tests show 5 defective modules in 20 tests. Paging down the data range, we may find some rows where sample testing for regular current got 5 defective modules, while some other rows where 5 defective modules occurred are cases where the electric current is irregular. To get more precise information out of our huge data set, we need to be able to efficiently count the number of times that 5 defective modules (or any other number of defective modules) occurred with either electric current level.

To analyze the electric current levels where in the simulations where sample testing of 20 modules got exactly 5 defective ones, let us first enter the number 5 into the cell D34. (Cell D34 will be our comparison cell that we can change if we want to count the number of occurrences of some other number of defective modules.) Next, into cell E36, we enter the formula

=IF(C36=$D$34,B36,"..")

30	FORMULAS		
31	B35. =IF(D2=0.5,1,0)		
32	C35. =D10		
33			
34		Current Irregular?	# Defective modules
35	Simulation table	1	9
36	0	1	7
37	0.001	0	1
38	0.002	1	11
39	0.003	1	13
40	0.004	0	2
41	0.005	0	3
42	0.006	1	8
43	0.007	0	2
44	0.008	0	2
45	0.009	0	3

(Data continues to row 1036)

Figure 2.6   Data table for simulation of 20 sample tests of electronic modules.

Then we copy the range E36 and paste it to the range E36:E1036. Now the cells in the E column have the value 1 in each data row where sample testing for a production run with irregular electric current get 5 defective modules, the value 0 in each data row where sample testing for a production run with regular electric current get 5 defective modules, and the value ".." in all other data rows. (Notice the importance of having absolute references to cell $D$34 in the above formulas, so that they do not change when they are copied.) With the cells in E36:E1036 acting as counters, we can count the number of times that 5 defective modules occurred in our simulation data by entering

=COUNT(E36:E1036)

into cell E28 (say). In this formula we are using the fact that Excel's COUNT function ignores cells that have non-numerical values (like ".." here). We can also use the same E cells to tell us how many times that 5 defective modules occurred for irregular, by entering the formula

$$=SUM(E36:E1036)$$

into cell E27. Here we are using the fact that Excel's SUM function also ignored non-numerical cells.

In a sample simulation data, the COUNT formula shows 5 defective modules occurred in 38 of our 1001 simulations, while the SUM formula shows that product runs for irregular were responsible for 6 of these 38 simulations where 5 defective modules occurred. Now, suppose that we have a new production run whose electric current is not known to us, but we learn that sample testing also got 5 defective modules. If our uncertainty about the electric current and defective modules for this production run are like those in our simulation data, then the sample testing should look like another draw from the population that gave us these 38 simulated sample tests which got 5 defective modules. Thus, given this sample testing information, we can estimate that the probability of irregular electric current is approximately 6/38 = 0.158. This result is computed by the formula =E27/E28 in cell E30 of Figure 2.6, which returns the same number that we could also get by the formula

$$=AVERAGE(E36:E1036).$$

This number 6/38 = 0.158 in cell E30 may be called our estimated conditional probability of a production run having irregular current given that the sample testing has got five defective modules in twenty 20 tested ones, based on our simulation data. In the notation of probability and decision-making theories, we often use the symbol "|" to denote the word "given", and mathematicians often use "≈" to denote the phrase "is approximately equal to". With this nota-

tion, the result of our simulation may be summarized by writing
P(Irregular Current | DefectiveModules=9) $\approx$ 13/68.

The conditional probability in cell E30 could also have been
computed by the formula =AVERAGE (E36:E1036), because Excel's AVERAGE function also ignores non-numerical cells. That is,
AVERAGE (E36:E1036) would return the value 6/38 = 0.158, because the range E36:E1036 contains 38 numerical values, of which
6 are ones and the rest are all zeros (see Figure 2.7).

	A	B	C	D	E	F	G	H
25	FORMULAS FROM RANGE A26:E1036				With Defective Modules=$D$34,			
26	B35. =IF(D2=0.5,1,0)				Frequency in Simulation Table:			
27	C35. =D10					6 Current Irregular		
28	E36. =IF(C36=$D$34,B36,"..")					38 All		
29	E36 copied to E36:E1036				P(Current Irregular|Defective Modules=$D$34)			
30	E27. =SUM(E36:E1036)				0.157895			
31	E28. =COUNT(E36:E1036)							
32	E30. =E27/E28							
33					Given Defective Modules =			
34		Current Irregular?	# Defective modules		5			
35	Simulation table	1	9		Current Irregular?			
36	0	1	7		..			
37	0.001	0	1		..			
38	0.002	1	11		..			
39	0.003	1	13		..			
40	0.004	0	2		..			
41	0.005	0	3		..			
42	0.006	1	8		..			
43	0.007	0	2		..			
44	0.008	0	2		..			
45	0.009	0	3		..			

Figure 2.7 Simulation data and analysis for electronic module
sample test.

2.4 SIMULATION FOR WHAT-IF ANALYSIS

What-if analysis is a useful tool for engineering decision making. Changing the 9 in cell D34 to other numbers, we can see the outcome frequencies for other numbers of defective units. But it would be helpful to make a table in which we can see together the results for all defective modules numbers from 0 to 20. Such a table can be easily made with Excel's data-table command (sometimes called a "what-if table" in other spreadsheet programs).

Here we will learn to use one form of data table called a column-input data table. The structure of a column-input data table is similar to the simulation data table that we made in the previous section.

To make our data table, we must begin by putting the output that we want to tabulate into one row, with space underneath to make the table. We want to tabulate the information contained in the frequency numbers in cells E27 and E28 of Figure 2.6, but there are not enough blank cells underneath these cells to make the data table there. So let us enter the formula =E27 into cell I34, to echo there the number of irregular currents that resulted in the given number of defective electronic modules in cell D34. Next let us enter the formula =E28-E27 into cell J34, to display there the number of regular currents that resulted in the given number of defective modules. Then to display the fraction of irregular among the total for the given number of defective electronic modules, let us enter the formula =I34/(I34+J34)into cell K34. This range I34:K34 will be the output range at the top of the data table. Underneath, the data table will tabulate the values of these cells as the parameter in cell D34 is adjusted from 9 to other values between 0 and 20.

The other values that we want to substitute into cell D34 must be listed in the column to the left of the output range, in the rows below it. So we must enter the numbers 0 to 20 into the cells from H36 to H56. (To do so quickly, first enter the number 0 into cell H36, and then we can select the range H36:H56 and use the command sequence Edit>Fill>Series, using the Series dialogue-box options: Columns, Linear, Step-Value 1, and Stop-Value 20.)

Now we select the range H34:K55 and use the command sequence Data>Table. When the "Row and Column Input" dialogue box comes up, we leave the "Row Input" entry blank, but we tab to the "Column Input" box and enter cell D34 as the Column Input Cell.

Following this Data>Table command, Excel will compute the data entries into the range I35:K55 as follows. For each row in this range, Excel first takes the value of the cell in column H (the leftmost column in our selected range H34:K55) and enters it into cell D34. Then Excel recalculates the whole spreadsheet. The new values of the output row at the top of the data table I34:K34 are then copied down into the corresponding (I, J, K) cells in this row. When all the cells in I35:K55 have been filled in this way, Excel restores the original contents of the input cell D34 (the value 9). Notice that the data in row 44 of the data table (I44:K44) is identical to the output above in I34:K34, because row 44 is based on the input value of 9 (from H44), which is the actual current value of cell D34 (as shown previously in Figure 2.7).

If you check the formulas in the cells from I35 to K55, you will find that they all share the special formula {=TABLE(,D34)}. The braces mean that this is an array formula, which Excel has entered into the whole range I35:K55 at once. (Excel will not let you change any one cell in an array; you have to change all or none. To emphasize that these cells together form an array, we have put a border around the data range I35:K55 in Figure 2.8, using a Format>Cells command.) The TABLE formula tells us that this range contains the data range of a data table that has no row-input cell but has D34 as its column-input cell. But recall that the whole range that we selected before invoking the Data>Table command also included one row above this data range and one column to the left of this data range. The column on the left side of the data table contains the alternative input values that are substituted one at a time into the designated column-input cell. The row at the top of the data table contains the output values that are tabulated as these alternative substitutions are done.

	H	I	J	K	L	M	N	O	P
32		Frequencies							
33		Current irregular	Current regular		P(CurrentIrregular|DefectiveModules)				
34	Defective Modules	6	32	0.157895		FORMULAS FROM RANGE H32:K35			
35	0	0	82	0		I34. =E27			
36	1	0	198	0		J34. =E28-E27			
37	2	1	205	0.004854		K34. =I34/(I34+J34)			
38	3	0	145	0		I35:K55. {=TABLE(,D34)}			
39	4	0	56	0					
40	5	6	32	0.157895					
41	6	8	8	0.5					
42	7	17	1	0.944444					
43	8	40	1	0.97561					
44	9	31	0	1					
45	10	60	0	1					
46	11	45	0	1					
47	12	28	0	1					
48	13	25	0	1					
49	14	9	0	1					
50	15	3	0	1					
51	16	0	0	#DIV/0!					
52	17	0	0	#DIV/0!					
53	18	0	0	#DIV/0!					
54	19	0	0	#DIV/0!					
55	20	0	0	#DIV/0!					

Figure 2.8 Simulation data table of results for different numbers of defective modules.

The conditional probability of any event A given some other event B, denoted by the formula P(A*B), is the probability that we would assign to this event A if we learned that the event B occurred. So the ratios in the K column of the data table give us estimates of the conditional probability of electronic current being irregular, given the total number of defective electronic modules found during sample testing of 20 modules. Notice how these conditional probabilities of being irregular increase from 0 to 1 as the given total number of defective modules.

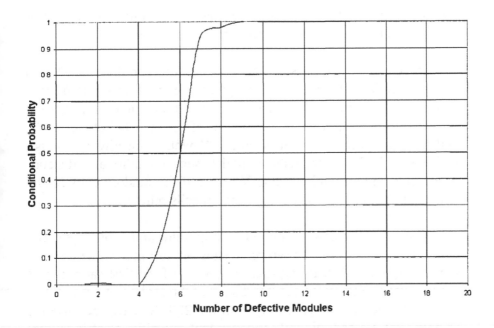

Figure 2.9 Plot of conditional probability of irregular electric current vs. number of defective modules.

Our simulation table gives us no data in Figure 2.8 for the extreme cases of 16 to 20 defective electronic modules. But it is obvious from the entries immediately above cell K51 that finding 16 or
more defective electronic modules in the 20 sample tests should
make us almost sure that the electronic current is irregular. The conditional probability of irregular electric current vs. number of defective modules is shown in Figure 2.9.

2.5 SIMULATING CONDITIONAL INDEPENDENCE

Consider again the spreadsheet model in Figure 2.5. In this simulation model, the results of the twenty tests calls in the cells B2:B21
are not independent, because they all depend on the random electronic in cell D2. But notice these results also depend on some independent random factors in cells A2:A21. In this case, we may say

that the results of the 20 sample tests are conditionally independent of each other when the electric current's level is given.

In general, when we say that some random variables, say X and Y, are conditionally independent given some other random variables, say Z, we mean that once you learned the value of Y, getting further information about X would not affect your beliefs about Y, and getting further information about Y would not affect your beliefs about X. In a spreadsheet model, such conditional independence holds among random cells X and Y if the random cells X and Y are not both influenced by any random cells other than Z.

Conditional independence is an important and subtle idea. Because the results of the 20 sample testing are not independent in our model (Figure 2.5), learning that the results of the first sample tests could cause use to revise our beliefs about the probability of a defective electronic module resulting from the 20th test. But because the sample tests are conditionally independent given the electric current level in this model, if we knew that the electric current is irregular then we would think that the probability of finding a defective module in the 20th test was 0.5, even if the sample testing had not found a defective module in any of the first 19 tests.

These concepts of conditional probability and conditional independence will be very important for describing what we do in our spreadsheet simulation models. With this terminology, the analysis of our model in this chapter can be summarized as follows:

The electric current may be either regular or irregular, with probabilities 0.7 and 0.3 respectively. Each of 20 tests may result in either a good module or a defective module. The results of 20 sample tests are conditionally independent of each other given the level of electric current. In each test, the conditional probability of a defective electronic module would be 0.5 given that the electric current is irregular, but the conditional probability of a defective module would be 0.1 given that the electric current is regular. Given this situation, we analyzed data from 1001 simulations of the model to estimate that the conditional probability that the electric current is irregular, given 5 defective modules found in the 20 sample tests, would be approximately 0.158.

2.6 PRODUCTION QUANTITY AND QUALITY MEASURES

It is always essential that defect rate types of systems measure the quality of the products and/or services produced by system processes. Frequently, it is also highly desirable to measure other aspects of the production or service process, including quantity measures of key variables such as production efficiency, production cycle times, and measures of flexibility involved in the production process. This short paper will suggest several methods that are useful for each.

Using C_{pk}

In Chapter 8, we will learn how to compute C_{pk} for a process. In principle, higher C_{pk}'s are better than lower Cpk's and because of it many companies have C_{pk} goals. For example, Motorola's Six-Sigma Quality initiative is essentially a goal of having C_{pk}'s greater than two for all of the relevant processes. Thus, one balanced scorecard measure might be to measure the C_{pk} of each process and then compute a weighted average of the C_{pk}'s to find a company average.

SPC Extent

It is universally clear that most companies that are attempting to achieve excellence are also attempting to get all of their processes under statistical control. Unless a process is under statistical control, to get quality, the company must inspect 100% of the output of the process. Thus, another potential balanced scorecard measure could be the percentage of the relevant processes that are under statistical control.

Process Steps

Good processes most frequently have fewer steps involved than do poorer processes. Fewer steps mean quicker, higher quality and more responsive processes. Reengineering of processes is targeted to removing steps that do not add value. Continually monitoring the number of total process steps in a system. the total number of non-value-adding steps, and the number of steps removed is another potential balanced scorecard type of process measurement that somewhat indirectly measures quality but also measures efficiency.

Other Quality-Related Measures

Many other indirect measures related to quality exist. Some of these are quite obvious and include:

> Percent scrap

> Percent rework

> Percent of labor dollars spent on inspection

Outgoing Product Quality Measures

A whole variety of measures exist for final products, some of which are made when prior to the product being shipped. some of which are customer reported. These include:

> Percent that are perfect when shipped

> Percent that works right the first time [out of the box]

> Percent of warranty claims

> Percent requiring service

> Expenditures on engineering changes

> Average time between service calls

Incoming Material Quality and Delivery Measures

It should be very evident that high quality incoming material from other vendors is prerequisite to smoothly functioning processes and high quality products. Thus, it is essential that incoming product be monitored. In businesses in which most, or all, of the material used in manufacture is outside the control of the business and solely dependent on outside vendors, incoming material quality could well be a potential balanced scorecard type of measure.

Several methods exist for ensuring the incoming quality of material. They include certifying vendors, demanding control charts for incoming product and sampling incoming product among others. Well-defined statistical methodologies exist for sampling incoming product when that is necessary using the MIL-STD-105E and MIL-STD-414 systems developed during World War II. Certified vendors are vendors who adhere to manufacturing practices that ensure the quality of their output and have demonstrated that fact to you. You have inspected their facilities, reviewed their SPC and quality practices and know that their processes are under control and meet your specification limits. Quality audits are regularly performed using the MIL-STD procedures to insure compliance.

Some potential global measures in this area are:

Percent of vendors that are certified

Percent of lots sampled lots that are rejected

Frequency of times production process problems or quality can be traced to incoming product quality

Additionally, if the manufacturing organization is a just-in-time (JIT) based organization, having the correct number of the correct parts arrive on time, every time is essential to a smoothly flowing production environment that is capable of meeting customer expectations.

Some potential measures in this area include:

Percent of deliveries that arrive within the prescribed incoming material time frame specifications

Percent of deliveries that have the correct amount of the correct material

Quantity-Related Measures

Many of the measures below are very traditional measurements used in manufacturing. The manager using them should, however, realize that the heart of the system, the area that needs the most analysis, are bottleneck activities since they drive the throughput of the manufacturing or service organization. Thus, although these measures are good in their place, from an operational perspective, measurements related to bottleneck activities are *infinitely* more valuable than simply monitoring every process and/or activity in a business.

Efficiency-Related Measures

Simple measures that measure the efficiency of a process or a system include measures related to a number of the process steps mentioned above. Other measures include the percent of labor hours spent on value-added operations and the ratio of direct labor to total labor for the business. Many other measures could be created that are variations on these themes.

Process Time-Related Measures

Cycle time is one of the most important variables in manufacturing. In typical long cycle times, non-value added activities form a high percentage of the cycle. Long cycle times create tremendous problems for manufacturers including the following:

High work-in-progress inventory - materials are sitting on the floor waiting to be processed

Make finished products inventory - when a company cannot make product quickly enough, it must create finished products inventory so customer demand can be quickly satisfied. This creates potential inventory holding costs and obsolescence costs.

Production changes - when goods take a long time to make, it is much more likely that changes will occur, including production amount, engineering changes, and actual changes to the process. All of these are expensive.

Complex systems - when parts are sitting out on the floor, complex inventory systems are needed to track them. These systems are expensive to build and maintain.

Uneven loading of work centers - when large batch sizes with long lead times are made, it is very difficult to balance the flow of materials through the plant system. World class manufacturers reduce the size of the production lot to improve the flow.

Inflexibility - long cycle times cause difficulties in responding to changing customer needs

Today's world class manufacturers shorten cycle times by reducing lot sizes and by employing synchronized production planning and control. JIT is the key to this effort. Visual in-plant inventory systems such as Kanban are primarily reactive and are no substitute to an effective JIT system. Measurements that are used to monitor process times include:

Manufacturing cycle time - this is typically done by continual recording actual cycle data into a tracking system or manually, by sampling the system periodically, and by using pragmatic methods (for example dividing the number of products produced by the work in process remaining). Simply tracking the average lot size can often be a key indicator.

D/P ratio - the ratio D/P = [Delivery Lead Time/Production Lead Time] in which delivery lead time is the amount of time is the lead time offered to customers and the production lead time is the amount of time required to make the product. When D is greater than P the product can be made to order. When D is less than P, the product must be made to stock. The lower the ratio, the more likely the customer will be happy because he has more flexibility. The trick with this ratio, is figuring out how to accurately measure D & P.

Setup times - with small lot sizes, setup time becomes critical. Typically work teams work together over a period of time to reduce setup times. The most typical measure of setup time is the standard deviation. Ideally, one would want the average setup time to be small, with a very low standard deviation. This would indicate that the production team can perform setups quickly with great precision, thus creating little disturbance to smooth product flow in the system.

Material availability - percent of time all parts needed are available; percent of job schedules that can be released for production are two reasonable measures

Distance of material movement during production - short distances; measured from engineering drawings; fork-lift mileage

Machine up-time - percentage of time machine is producing parts; percentage of time machine is down for repair

Customer service time - time it takes from the time the order is received until the customer receives the product

Production Flexibility Related Measures

Production flexibility is highly desirable in today's customer driven marketplace. Companies with short production cycles are much more flexible than those with long cycle times. The commonality of components and subassemblies, the degree of process commonality between products, and the number of subassemblies (fewer = better) are all important characteristics of flexible processes.

In addition, production flexibility has another major component--workers who are able to deliver multiple skills, solve problems, learn new skills continuously, and work independently.

Potential measures that indicate the flexibility of a company's manufacturing facility include:

Number of different parts in a bill of materials (smaller = better)

Percentage of standard, common, and unique parts in the bill of materials (higher = better)

The number of production processes (fewer = better)

The number of new products introduced each year and the time taken to launch them (more new products: more = better; time to launch average: smaller = better)

Cross-training of production personnel (more cross-trained = better)

Comparison of production output and production capacity (running close to 100% of capacity across all processes = better)

Several possible measures exist for measuring the quality, the quantity, and the production capabilities for a company. What needs to be used are a relatively few measures that really form key indicators for a manager of what is happening in the plant or company. For companies that are being forced to change from traditional manufacturing and service organizations to ones that suddenly find themselves engaged in simultaneous battles to achieve world-class quality, needing to meet continually changing customer requirements, and having to continuously drive down their costs, must have measures to know where they are at on all fronts at all times. Without these measures, they will most certainly fail. Picking a SMALL number of SIGNIFICANT measures that truly measure the most important variables, collecting them accurately and systematically, and displaying them simply can form the difference between success and failure in the battle for competitive success.

2.7 SUMMARY

This chapter has focused on an example in which we want to learn about some unknown quantity (the electric current level) by observing other events (the successes and failures in various sample tests) that are influenced by this unknown quantity. We have analyzed this problem using spreadsheet simulation models in which one random cell simulates the unknown quantity and other cells simulate the observations that depend on the unknown quantity.

In the context of this problem, we have introduced some basic ideas of probability and some basic techniques of spreadsheet modeling. Probability ideas introduced here include: prior and conditional probabilities P(A*B), independence, conditional independence, uniform and normal probability distributions, cumulative probabilities, and (inverse) cumulative probability charts. Excel functions introduced here include: RAND, IF, COUNT, SUM, AVERAGE, PERCENTILE, and FREQUENCY. We have described how to get information about these and other technical functions in Excel, by the using of the insert-function dialogue box. Other basic spreadsheet techniques used in this chapter include: absolute ($) and relative references in formulas, simulation tables, column-input data tables, filled series, the paste-special-values command, and the basic XY-Scatter chart.

To compute conditional probabilities from large tables of simulation data, we have introduced a formula-filtering technique in which a column is filled with IF formulas that extract information from a simulation table, returning a non-numerical value ("..") in data rows that do not match our criterion. The information extracted in such columns has been summarized using statistical functions like COUNT, SUM, AVERAGE, FREQUENCY, and PERCENTILE, which are designed to ignore non-numerical entries.

The role of engineering judgment is emphasized when dealing with practical problems; engineering judgments are often necessary irrespective of theoretical sophistication. However, the proper place and role for engineering judgments are delineated within the overall analysis of uncertainty and its effects on decision and risk. As shown in this chapter, consistent with a probabilistic approach, engineering judgments have to be expressed in probability or statistical terms; wherever expert judgments are expressed in conventional or deterministic terms, they have to be translated into appropriate probabilistic terms. Methods for these purposes are presented in the next chapter.

REFERENCES

Anon. (1989), Risk Management Concepts and Guidance, Defense Systems Management College, Fort Belvoir VA.

Batson, R. G. (1987). "Critical Path Acceleration and Simulation in Aircraft Technology Planning," IEEE Transactions on Engineering Management, Vol. EM-34, No. 4, November, pp. 244-251.

Batson, R. G. and R. M. Love (1988). "Risk Assessment Approach to Transport Aircraft Technology Assessment," AIAA Journal of Aircraft, Vol. 25, No. 2, February, PP. 99-105.

Bell, T. E., ed. (1989). "Special Report: Managing Risk in Large Complex Systems," IEEE Spectrum, June, pp. 21-52.

Beroggi, G. E. G., and W. A. Wallace (1994). "Operational Risk Management: A New Paradigm for Decision Making," IEEE Transactions on Systems, Man, and Cybernetics, Vol. 24, No. 10, October, pp. 1450-1457.

Black, R. and J. Wilder (1979). "Fitting a Beta Distribution from Moments," Memorandum, Grumman, PDM-OP-79-115.

Book, S. A. and P. H. Young (1992). "Applying Results of Technical-Risk Assessment to Generate a Statistical Distribution of Total System Cost," AIAA 1992 Aerospace Design Conference, Irvine CA, 3-6 February.

Chapman, C. B. (1979). "Large Engineering Project Risk Analysis," IEEE Transactions on Engineering Management, Vol. EM-26, No. 3, August, pp. 78-86.

Chiu, L. and T. E. Gear (1979). "An Application and Case History of a Dynamic R&D Portfolio Selection Model," IEEE Transactions on Engineering Management, Vol. EM-26, No. 1, February, pp. 2-7.

Cullingford, M. C. (1984). "International Status of Application of Probabilistic Risk Assessment," Risk & Benefits of Energy Systems: Proceedings of an International Symposium, Vienna Austria, IAEA-SM-273/54, pp. 475-478.

Dean, E. B. (1993). " Correlation, Cost Risk, and Geometry," Proceedings of the Fifteenth Annual Conference of the International Society of Parametric Analysts, San Francisco CA, 1-4 June.

Degarmo, E. P., Sullivan, W. G., Bontadelli, J. A., and Wicks, E. M. (1997), Engineering Economy, Tenth Edition, Prentice Hall, Upper Saddle River, New Jersey.

Dienemann, P. F. (1966), Estimating Cost Uncertainty Using Monte Carlo Techniques, The Rand Corporation, Santa Monica CA, January, RM-4854-PR.

Dodson, E. N. (1993), Analytic Techniques for Risk Analysis of High-Technology Programs, General Research Corporation, RM-2590.

Fairbairn, R. (1990), "A Method for Simulating Partial Dependence in Obtaining Cost Probability Distributions," Journal of Parametrics, Vol. X, No. 3, October, pp. 17-44.

Garvey, P. R. and A. E. Taub (1992), "A Joint Probability Model for Cost and Schedule Uncertainties," 26th Annual Department of Defense Cost Analysis Symposium, September.

Greer, W. S., Jr. and S. S. Liao (1986). "An Analysis of Risk and Return in the Defense Market: Its Impact on Weapon System Competition," Management Science, Vol. 32, No. 10, October, pp. 1259-1273.

Hazelrigg, G. A., Jr. and F. L. Huband (1985). "RADSIM - A Methodology for Large-Scale R&D Program Assessment," IEEE Transactions on Engineering Management, Vol. EM-32, No. 3, August, pp. 106-115.

Henley, E. J. and Kumamoto, H. (1992), "Probabilistic Risk Assessment," IEEE Press, Piscataway, NJ.

Hertz, D. B. (1979). "Risk Analysis in Capital Investment," Harvard Business Review, September/October, pp. 169-181.

Honour, E. C. (1994). "Risk Management by Cost Impact," Proceedings of the Fourth Annual International Symposium of the National Council of Systems Engineering, Vol. 1, San Jose CA,10-12 August, pp. 23-28.

Hutzler, W. P., J. R. Nelson, R. Y. Pei, and C. M. Francisco (1985). "Nonnuclear Air-to-Surface Ordnance for the Future: An Approach to Propulsion Technology Risk Assessment," Technological Forecasting and Social Change, Vol. 27, pp. 197-227.

Kaplan, S. and B. J. Garrick (1981). "On the Quantitative Definition of Risk," Risk Analysis, Vol. 1, No. 1, pp. 11- 27.

Keeney, R. L. and D. von Winterfeldt (1991). "Eliciting Probabilities from Experts in Complex Technical Problems," IEEE Transactions on Engineering Management, Vol. 38, No. 3, August, pp. 191-201.

Markowitz, H. M. (1959). Portfolio Selection: Efficient Diversification of Investment, 2nd. ed., John Wiley & Sons, New York NY, reprinted in 1991 by Basil Blackwell, Cambridge MA.

McKim, R. A. (1993). "Neural Networks and the Identification and Estimation of Risk," Transactions of the 37th Annual Meeting of the American Association of Cost Engineers, Dearborn MI, 11- 14 July, P.5.1-P.5.10.

Ock, J. H. (1996). "Activity Duration Quantification Under Uncertainty: Fuzzy Set Theory Application." Cost Engineering, Vol. 38, No. 1, pp. 26-30.

Quirk, J., M. Olson, H. Habib-Agahi, and G. Fox (1989). "Uncertainty and Leontief Systems: an Application to the Selection of Space Station System Designs," Management Science, Vol.35, No. 5, May, pp. 585-596.

Rowe, W. D. (1994). "Understanding Uncertainty," Risk Analysis, Vol. 14, No. 5, pp. 743-750.

Savvides, S. (1994). "Risk Analysis in Investment Appraisal," Project Appraisal, Vol. 9, No. 1, March, pp. 3-18.

Sholtis, J. A., Jr. (1993). "Promise Assessment: A Corollary to Risk Assesment for Characterizing Benefits," Tenth Symposium on Space Nuclear Power and Propulsion, Alburquerque NM, American Institute of Physics Conference Proceedings 271, Part 1, pp. 423-427.

Shumskas, A. F. (1992). "Software Risk Mitigation," in Schulmeyer, G. G. and J. I. McManus, ed., Total Quality Management for Software, Van Nostrand Reinhold, New York NY.

Skjong, R. and J. Lerim (1988). "Economic Risk of Offshore Field Development," Transactions of the American Association of Cost Engineers, New York NY, pp. J.3.1-J.3.9.

Thomsett, R. (1992), "The Indiana Jones School of Risk Management," American Programmer, vol. 5, no. 7, September 1992, pp. 10-18.

Timson, F. S. (1968), "Measurement of Technical Performance in Weapon System Development Programs: A Subjective Probability Approach," The Rand Corporation, Memorandum RM-5207-ARPA.

Williams, T. (1995). "A Classified Bibliography of Recent Research Relating to Project Risk Management," European Journal of Operations Research, Vol. 85, pp.18-38.

3

Decision Analysis Involving Continuous Uncertain Variables

Uncertainty about continuous uncertain variables is pervasive in many engineering decisions. Here the engineer is confronted with a broader scope of uncertainty, involving several unknown quantities, some of which could have infinitely many possible values. The primary responsibility of an engineer is to make decisions based on predictions and information that invariably contain these uncertainties. Through engineering decision analysis, uncertainties can be modeled and assessed properly, and their effects on a given decision accounted for systematically. In this chapter, we will discuss methods and spreadsheet simulation models for analyzing such complex uncertainties.

3.1 CASE STUDY: A BIOENGINEERING FIRM

Penn-Technology Inc. (PTI) is a bioengineering firm, founded 5 years ago by a team of biologists. A venture capitalist has funded PTI's research during most of this time, in exchange for a controlling interest in the firm. PTI's research during this period has been focused on the problem of developing a bacterium that can synthesize HP-273, a protein with great pharmacological and medical potential.

This year, using new methods of genetic engineering, PTI has at last succeeded in creating a bacterium that produces synthetic HP-273 protein.

PTI can now get a patent for this bacterium, but the patented bacterium will be worthless unless the FDA approves the synthetic protein that it produces for pharmacological use. The process of getting FDA approval has two stages. The first stage involves setting up a pilot plant and testing the synthetic protein on animal subjects, to get a preliminary certification of safety which is required before any human testing can be done. The second stage involves an extensive medical study for safety and effectiveness, using human subjects.

PTI's engineers believe that the cost of the first stage is equally likely to be above or below $9 million, has probability 0.25 of being below $7 million, and has probability 0.25 of being above $11 million. The probability of successfully getting a preliminary certification of safety at the end of this first stage is 0.3.

The cost of taking the synthetic protein though the second-stage extensive medical study has a net present value that is equally likely to be above or below $20 million, has probability 0.25 of being below $16 million, and has probability 0.25 of being above $24 million. Given first-stage certification, if PTI takes the synthetic HP-273 protein through this second stage then the conditional probability of getting full FDA approval for its pharmacological use is 0.6.

If PTI gets full FDA approval for pharmacological use of the synthetic HP-273 protein, then PTI will sell its patent for producing the synthetic HP-273 protein to a major pharmaceutical company. Given full FDA approval, the returns to PTI from selling its patent would have a present value that is equally likely to be above or below $120 million, has probability 0.25 of being below $90 million, and has probability 0.25 of being above $160 million.

PTI's management has been operating on the assumption that they would be solely responsible for the development of this HP-273 synthetic protein throughout this process, which we will call Plan A. But there are two other alternatives that the owner wants to consider:

Plan B: PTI has received an offer to sell its HP-273 patent for $4 million now, before undertaking the any costs in the first stage of the approval process.

Plan C: If PTI succeeds in getting preliminary FDA certification at the end of the first stage, then it could to sell a 50% share of the HP-273 project to other investors for $25 million.

Here we need to estimate the expected value and standard deviation of PTI's profits under Plan A, Plan B and Plan C. Then make a cumulative risk profile for each plan. If the owner of PTI wants to maximize the expected value of profits, then what plan should we recommend, and what is the lowest price at which PTI should consider selling its HP-273 patent now? Compute the probability that the stage 1 cost will be between $8 and $12 million. Also compute the probability that it will be more than $12 million.

Uncertainty about numbers is pervasive in all engineering decisions. How many units of a proposed new product will we manufacture in the year when it is introduced? What is the defective rate for a new production process? What will be the total warranty cost for this new products? Each of these number is an unknown quantity. If the profit or payoff from a proposed strategy depends on such unknown quantities, then we cannot compute this payoff without making some prediction of these unknown quantities.

Chapter 2 introduced a simple decision problem with one unknown quantity that had limited number of possible values. In this chapter, the engineering analyst is confronted a broader scope of uncertainty, involving several unknown quantities, some of which could have infinitely many possible values. In this chapter, we will develop methods for analyzing such complex situations.

The way to make this complex task tractable is to assume that our beliefs about the unknown quantity can be reasonably approximated by some probability distribution in one of the mathematical families of probability distributions that have been studied by statisticians. Each of these families includes a variety of different probability distributions that are characterized by a few parameters. So our task is, first to pick an appropriate family of probability distributions, and then to assess values for the corresponding parameters that can fit our actual beliefs about the unknown quantity. The most commonly used families of probability distributions typically have only two or three parameters to assess.

This chapter introduces some of the best-known families of probability distributions, each of which is considered particularly appropriate for certain kinds of unknown quantities. But in general practice, the *Normal* and *Lognormal* distributions are widely recognized as applicable in the broadest range of situations. In this chapter, we will give most careful consideration to these two families.

3.2 NORMAL DISTRIBUTIONS

The central limit theorem tells us that, if we make a large number of independent samples from virtually any probability distribution, our sample average will have a probability distribution that is approximately Normal. This remarkable theorem assures us that there will be many unknowns in the real world that are well described by Normal distributions. In fact, a Normal probability distribution will be a good fit for any unknown quantity that can be expressed as the sum of many small contributions that are independently drawn from some fixed probability distribution. For example, if our uncertainty about each day's production is the same as any other day and is independent of all other days' production, and each day's production will be small relative to the whole year, then total annual production should be Normal random variable.

The Normal probability distributions are a two-parameter family that is indexed by the mean and standard deviation. That is, for any number μ and any positive number σ, there is a Normal probability distribution that has mean μ and standard deviation σ.

A linear transformation of any Normal random variable is also Normal. That is, if a random variable X has a Normal probability distribution with mean μ and standard deviation σ, and if c and d are any two nonrandom numbers, then the random variable c*X+d also has a Normal probability distribution, with mean c*μ+d and standard deviation |c|*σ.

Every number (positive or negative) is a possible value of a Normal random variable. Of course there are infinitely many numbers, and so no single number gets positive probability under a Normal distribution. Instead, we can only talk meaningfully about intervals of numbers having positive probability. In the language of probability theory, such distributions that do not put positive probability on any single value are called continuous.

Excel provides several functions for dealing with Normal distributions. Here we will use two: NORMINV and NORMSDIST. The function NORMINV will be much more important for computer simulation.

The cumulative probability below any value in a Normal distribution depends on how many standard deviations this value is above or below the mean, according to the Excel function NORMSDIST. That is, if X is any Normal random variable with mean μ and standard deviation σ, and w is any number, then

$$P(X<w) = NORMSDIST((w-\mu)/F)$$

If we want to know the probability that X, a Normal random variable with mean μ and standard deviation σ, is between some pair of numbers w and y, where w < y, then we can use the formula

$$P(w < X < y) = P(X < y) - P(X < w)$$

$$= NORMSDIST((y-\mu)/F) - NORMSDIST((w-\mu)/F)$$

as illustrated by cell F3 in Figure 3.1.

The Normal distribution has a symmetry around its mean so that, for any number w,

$$NORMSDIST(-w) = 1-NORMSDIST(w)$$

That is, a Normal random variable with mean μ and standard deviation F is as likely to be less than μ-w*σ as it is to be greater than μ+w*σ.

Some special values of the NORMSDIST function worth noting are NORMSDIST(0) = 0.5, NORMSDIST(0.675) = 0.75, NORMSDIST(1) = 0.841, NORMSDIST(1.96) = 0.975, NORMSDIST(3) = 0.99865. These cumulative probabilities imply that, when X is any Normal random variable with mean μ and standard deviation σ,

$$P(X < \mu) = P(X > \mu) = 0.5$$
$$P(X < \mu-0.675*\sigma) = 0.25$$
$$P(X < \mu+0.675*\sigma) = 0.75$$
$$P(\mu-F < X < \mu+\sigma) = 0.683$$
$$P(\mu-1.96*\sigma < X < \mu+1.96*\sigma) = 0.95$$
$$P(\mu-3*\sigma < X < \mu+3*\sigma) = 0.9973$$

The Excel function NORMINV is the inverse cumulative-probability function for Normal probability distributions. That is, if

X is any Normal random variable with mean μ and standard deviation σ, then

$$P(X < NORMINV(q,μ, σ)) = q$$

Here the NORMINV parameters q, μ, and σ can be any three numbers such that σ>0 and 0<q<1. Using NORMINV, we can make an inverse cumulative distribution chart by plotting probability numbers from 0 to 1 against the corresponding NORMINV(probability,μ, σ) values, as shown in Figure 3.1 (in rows 16-30) for the values μ=28 and σ=5.

Then the probability density can be estimated in column C by entering into cell C7 the formula

$$=(A8-A6)/(B8-B6)$$

and then copying cell C7 to the range C7:C105. (This formula in the C column is not applied in the top and bottom rows 6 and 106, because it needs to refer to A and B values in the rows above and below the current row.) In this C7 formula, the denominator B8-B6 is the length of an interval from a value (12.55 in B6) that is slightly less than B7 to a value (17.73 in B8) that is slightly greater than B7 (which is 15.53). The numerator A8-A6 is the probability of the random variable being in this short interval, because A8 (0.02) is the probability of the random variable being less than B8 (17.73), while A6 (0.001) is the probability of the random variable being less than B6 (12.55). So (A8-A6)/(B8-B6) is the probability in a short interval around B7 divided by the length of this interval, and so it can be used to approximate the probability density of this random variable at B7.

The probability density chart shown in Figure 3.1 (in rows 33 to 47) is plotted from the values in B7:B105 and the estimated densities in C7:C105. The probability chart of a Normal distribution is often called a bell-shaped curve. The probability density function provides a useful way to visualize a random variable's probability distribution because, for any interval of possible values, the probability of the interval is the area under the probability density curve over this interval.

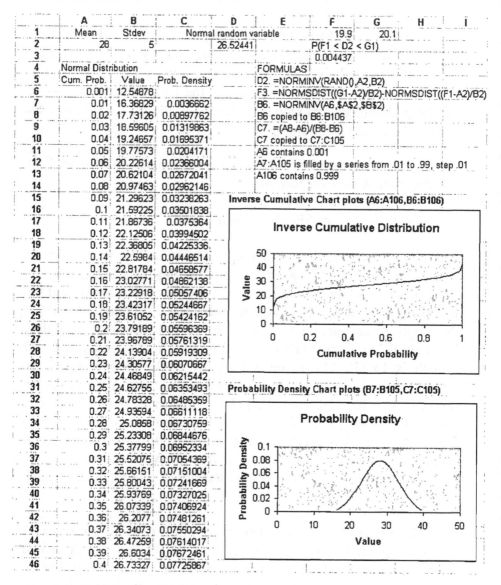

Figure 3.1 The Normal distribution.

Recall from Chapter 2 that we can simulate a random variable with any probability distribution by applying a RAND() value as input into the inverse cumulative-probability function. Thus, we can simulate a Normal random variable with mean μ and standard deviation by the formula

$$\text{NORMINV}(\text{RAND}(),\mu,\sigma)$$

as shown in cell D2 of Figure 3.1, where the mean μ is in cell A2 and the standard deviation F is in cell B2.

The probability density function provides a useful way to visualize a random variable's probability distribution because, for any interval of possible values, the probability of the interval is the area under the probability density curve over this interval. Excel provides a rather complicated formula for computing Normal probability-density functions [NORMDIST(C,μ,F,FALSE)]. The inverse-cumulative function NORMINV can help us to simulate such probability density functions.

A probability density function for a random variable at any possible value is actually defined mathematically to be the slope of the cumulative probability function at that value. So when a mathematician says that a random variable X has a probability density $f(y)$ at some value y, this statement means that, for any small interval around y, the probability of X being in that interval is approximately equal to the length of the interval multiplied by the density $f(y)$. According to Figure 3.1, for example, a Normal random variable with mean 28 and standard deviation 5 has, at the value 20, a probability density of approximately 0.024. So we may infer that the probability of such a random variable being between 19.9 and 20.1 (an interval of length 0.2 that includes the value 20) is approximately 0.2*0.024 = 0.0048.

To make a chart of probability densities for a given distribution, let us begin by filling a range of cells in column A with probability values that go from 0 to 1. For the spreadsheet of Figure 3.1, such probability values from 0 to 1 with step size 0.01 are filled into the range A6:A106, except that the values 0 and 1 at the top and bottom of this range are changed to 0.001 and 0.999, because NORMINV and many other inverse cumulative functions are not defined at the probabilities 0 and 1. Next, in column B, let us compute the corresponding inverse-cumulative values. So for the spreadsheet in Figure 3.1, with the mean in cell B2 and the standard deviation in cell C2, we enter in cell B6 the formula

$$=NORMINV(A6,\$A\$2,\$B\$2)$$

and we copy B6 to cells B6:B106.

Normal random variables can take positive or negative values, and their probability of going negative can be non-negligible when the mean is less than twice the standard deviation. But for there are many unknown quantities, such as product reliability, where a negative value would be nonsensical. For such unknowns, it is more appropriate to use a probability distribution in which only the positive numbers are possible values. The statistical literature features several families of such distributions, of which the most widely used in engineering decision analysis is the family of Lognormal probability distributions.

3.3 LOGNORMAL DISTRIBUTIONS

By definition, a random variable **Y** is a Lognormal random variable if its natural logarithm LN(Y) is a Normal random variable. Because the EXP function is the inverse of LN, a Lognormal random variable can be equivalently defined as any random variable that can be computed by applying the EXP function to a Normal random variable. Thus, for any number m and any positive number s, the formula

$$=EXP(NORMINV(RAND(),m,s))$$

is a Lognormal random variable. The parameter m and s are called the log-mean and log-standard-deviation of the random variable. The log-mean and the log-standard-deviation are not the mean and standard deviation of this Lognormal random variable, instead they are the mean and standard deviation of its natural logarithm. In general, the Lognormal random variables have a family of probability distributions that can be parameterized by their log-mean and log-standard-deviation. If you multiply a Lognormal random variable with log-mean m and log-standard-deviation by a (nonrandom) positive constant c, the result is another Lognormal random variable that has log-mean m+LN(c) and the same log-standard-deviation s.

Lognormal distribution can be used to model, for example, the physics-of-failure process in reliability engineering. The precise model that leads to a lognormal distribution is called a multiplicative growth or degradation model. At any instant of time, the process undergoes a random increase of degradation that is proportional to its present state. The multiplicative effect of all these random (and assumed independent) growths or degradations builds up to failure. To see why, let Y denote the degradation ratio for the value of some material strength over the next 30 years. That is, Y is the unknown value of this material strength 30 years from today divided by the material strength's value today. Here our system is designed for 30 years. For each integer i from 1 to 30, let X_i denote the degradation ratio of this material-strength degradation during the i'th year of the next 30 years. So

$$Y = X_1 * X_2 * ... * X_{30}$$

Taking the natural logarithm of both sides, we get

$$LN(Y) = LN(X_1 * X_2 * ... * X_{52}) = LN(X_1) + LN(X_2) + ... LN(X_{30})$$

because LN converts multiplication to addition. Now suppose that the degradation ratio each year is drawn independently from the same probability distribution. Suppose also that there cannot be a large change in any one year, that is, the degradation ratio X_i in any one year must be close to 1. Then $LN(X_1)$, ..., $LN(X_{30})$ will be independent random variables, each drawn from the same distribution on values close to $LN(1) = 0$. But recall from the beginning of Section 3.2 that a Normal probability distribution will be a good fit for any unknown quantity that can be expressed as the sum of many small (near 0) contributions that are independently drawn from a fixed probability distribution. So $LN(Y) = LN(X_1) + LN(X_2) + ... LN(X_{30})$ must be an approximately Normal random variable, which implies that the degradation ratio Y itself must be an approximately Lognormal random variable. Furthermore, multiplying this degradation ratio Y by the known current value of this material strength, we find that this material strength's value 30 years from today must also be an approximately Lognormal random variable.

This result is illustrated in Figure 3.2, which illustrates a material strength whose value can change by up to 5% on in any given year. To be specific, this model assumes that the degradation ratio of the material strength value is drawn from a uniform probability distribution over the interval from 0.90 to 1.00. These degradation ratios for the 30 years of the system design life are simulated by entering the formula =0.90+0.10*RAND() into each cell in the range A8:A37. Supposing that the initial value of this material strength is 1, cell B5 is given the value 1, the formula =B7*A8 is entered into cell B8, and then cell B8 is copied to B8:B37. The result is that cells B7:B57 simulate the value of this material strength at the end of each of the 30 years of the design life. The degradation ratio for the whole 30-years design life (which is also the value at the end of the design life if the initial value was 1) is echoed in cell E7 by the formula =B37. Because this annual degradation ratio is the multiplicative product of many independent random ratios, each of which is close to 1, we know that it must be an approximately Lognormal random variable.

Knowing that the annual degradation ratio in cell E7 of Figure 3.2 is a Lognormal random variable, we may now ask which Lognormal distribution is the probability distribution for this random variable. Recall that Lognormal probability distributions can be parameterized by their log-mean and log-standard-deviation, which are respectively the mean and standard deviation of the natural logarithm of the random variable. To estimate these parameters, we can compute the sample mean and standard deviation of many independent recalculations of LN(E7), by entering the formula =LN(E7) into a cell that serves as the model output at the top of a large simulation table (not shown in the figure). In this way (using a few thousand simulations), one can show that the log-mean is approximately −1.555 and the log-standard-deviation is approximately 0.166. That is, the formula

$$\text{=EXP(NORMINV(RAND(), -1.555, 0.166))}$$

should yield a random variable that has approximately the same probability distribution as the random 30-years degradation ratio in cell E7. Such a formula has been entered into cell F12 of Figure 3.2.

Of course the realized value of cells E7 and F12 are very different in Figure 3.2, because they are independent random variables. To show that they have the same probability distribution, we should make a simulation table that records several hundred recalculations of E7 and F12 (not shown in the figure). Then these simulated values of E7 and F12 should be sorted separately and plotted against the percentile index, to yield a chart that estimates the inverse cumulative probability distributions of these two random variables. These estimated inverse cumulative probability curves for E7 and F12 should be very close to each other.

We have argued in this section that the growth ratio of a financial investment over some period of time might be naturally assumed to be a Lognormal random variable. We have noted that this assumption then implies that the value of the investment as any particular date in the future would also be a Lognormal random variable (because the future value is the known current value multiplied by the Lognormal growth ratio between now and that future date). But this assumption also implies that logarithmic rates of return over any interval of time are Normal random variables, because the natural logarithm of a Lognormal random variable is a Normal random variable.

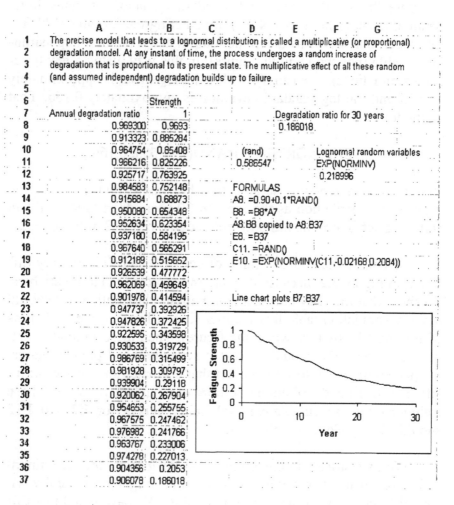

Figure 3.2   The Lognormal distribution.

To remember this distinction, notice that growth rates can be positive or negative, and Normal random variables can be positive or negative. But Lognormal random variables are never negative, and a financial asset's value at any point in time cannot be negative, and its growth ratio over any interval of time also cannot be negative.

3.4 SELECTION BETWEEN NORMAL AND LOGNORMAL DISTRIBUTIONS

Many probability distributions can be characterized by specifying their value at the 0.25, 0.50, and 0.75 cumulative-probability levels, which we may call the quartile boundary points for the distribution, denoted by Q_1, Q_2, and Q_3 respectively. That is, we may say that Q_1, Q_2, and Q_3 are the quartile boundary points for an unknown quantity X if

$$P(X < Q_1) = 0.25, P(X < Q_2) = 0.50, \text{ and } P(X < Q_3) = 0.75$$

These three boundary points divide the number line into four separate quartiles that are equally likely to include the actual value of X. The first quartile is the set of all numbers less than Q_1, the second quartile is interval between Q_1 and Q_2, the third quartile is the interval between Q_2 and Q_3, and the fourth quartile is the set of numbers greater than Q_3. Assuming that X is a continuous random variable, each of these quartiles has the same probability 1/4 of including the actual value of X. The second quartile boundary point Q_2 is also called the median of the distribution of X.

Selection of Normal Distribution

In the special case where the quartile boundary points satisfy the following equal-difference condition

$$Q_3 - Q_2 = Q_2 - Q_1,$$

this random variable is a Normal random variable with mean $\mu = Q_2$ and standard deviation $F = (Q_3 - Q_1)/1.349$.

In the case study, PTI's engineers believe that the cost of the first stage is equally likely to be above or below $9 million, has probability 0.25 of being below $7 million, and has probability 0.25 of being above $11 million. The quartile boundary points for development cost assessed by PTI engineers are $7, $9, and $11 million, which happen to satisfy the equal-difference condition that characterizes the Normal distributions (11-9 = 9-7). For a Normal distribution, the quartile boundary points Q_1, Q_2, Q_3 depend on the mean μ and standard deviation σ by the formulas

$$Q_1 = \mu\text{-}0.675*\sigma, Q_2 = \mu, Q_3 = \mu + 0.675*\sigma.$$

So with these assessed quartiles, the mean and standard deviation of the development cost must be

$$\mu = Q_2 = 9$$

and

$$\sigma = (Q_3 - Q_2)/0.675 = (11\text{-}9)/0.675 = 2.96$$

Thus, the development cost in this case could be simulated equally well by the formula NORMINV(RAND(), 9, 2.96).

Similarly, the cost of taking the synthetic protein though the second-stage extensive medical study has a net present value that is equally likely to be above or below $20 million, has probability 0.25 of being below $16 million, and has probability 0.25 of being above $24 million. The quartile boundary points for development cost assessed by PTI engineers are $16, $20, and $24 million, which also satisfy the equal-difference condition that characterizes the Normal distributions (24-20 = 20-16). So with these assessed quartiles, the mean and standard deviation of the development cost must be $\mu = Q_2 = 20$ and $\sigma = (Q_3 - Q_2)/0.675 = (24\text{-}20)/0.675 = 5.926$. Thus, the development cost in this case could be simulated equally well by the formula NORMINV(RAND(), 20, 5.926).

Selection of Lognormal Distribution

On the other hand, in the special case where the quartile boundary points satisfy the following equal-ratio condition

$$Q_3/Q_1 = Q_2/Q_1,$$

the random variable is a Lognormal random variable with log-mean

$$m = LN(Q_2) \text{ and}$$

log-standard-deviation

$$\sigma = (LN(Q_3)/LN(Q_1))/1.349.$$

In the case study, given full FDA approval, the returns to PTI from selling its patent would have a present value that is equally likely to be above or below $120 million, has probability .25 of being below $90 million, and has probability .25 of being above $160 million. The quartile boundary points for total market value assessed by the business-marketing manager are 90, 120, and 160 $million, which happen to satisfy the equal-ratio condition that characterizes the Lognormal distributions among the Generalized-Lognormals (120/90 =160/120). For a Lognormal distribution with log-mean m and log-standard-deviation s, the quartile boundary points Q_1, Q_2, Q_3 depend on the log-mean and log-standard-deviation by the formulas

$$LN(Q_1) = m-0.6745*s, \ LN(Q_2) = m, \ LN(Q_3) = m + 0.6745*s.$$

So with these assessed quartiles, the log-mean and log-standard-deviation of the total market value must be m = $LN(Q_2)$ = LN(120) = 4.79 and s = (LN(160)-LN(120))/0.6745 = (LN(120)-LN(90))/0.6745 = 0.4265. Thus, the total market value in this case could be simulated equally well by the formula EXP(NORMINV(RAND(),4.7875,0.4265)).

3.5 DECISION TREE ANALYSIS

An engineering decision may not be based solely on the available information. If time permits, additional information may be collected prior to the final selection among the feasible design alternatives. In engineering problems, additional data could take the form of research, analysis, development testing, field performance data, etc. Since there could be a variety of such schemes for collecting additional information, this would further expand the spectrum of alternatives. We will illustrated the expansion of spectrum of alternatives using the case study of the decision-making problem in the Section 3.1.

For general engineering decision problems, a framework for systematic analysis is required. Specifically, the decision analysis should at least include the following components:

1. A list of all feasible alternatives, including the acquisition of additional information, if appropriate;
2. A list of all possible outcomes associated with each alternatives;
3. An estimation of the probability associated with each possible outcome;
4. An evaluation of the consequences associated with each combination of alternative and outcome;
5. The criterion for engineering decision;
6. A systematic evaluation of all alternatives.

The various components of an engineering decision problem may be integrated into a formal layout in the form of a *decision tree*, consisting of the sequence of decisions: a list of feasible alternatives; the possible outcome associated with each alternative; the corresponding probability assignments; and monetary consequence and utility evaluations. In the other words, the decision tree integrates the relevant components of the decision analysis in a systematic manner suitable for analytical evaluation of the optimal alternative. Probability models of engineering analysis and design may be used to estimate the relative likelihood of the possible outcomes, and appropriate value or utility models evaluate the relative desirability of each consequence.

Decision trees model sequential decision problems under uncertainty. A decision tree graphically describes the decisions to be made, the events that may occur, and the outcomes associated with combinations of decisions and events. Probabilities are assigned to the events, and values are determined for each outcome. A major goal of the analysis is to determine the best decisions. Decision tree models include such concepts as nodes, branches, terminal values, strategy, payoff distribution, certainty equivalent, and the rollback method.

Nodes and Branches

Decision trees have three kinds of nodes and two kinds of branches. A decision node is a point where a choice must be made; it is shown as a square. The branches extending from a decision node are decision branches, each branch representing one of the possible alternatives or courses of action available at that point. The set of alternatives must be mutually exclusive (if one is chosen, the others cannot be chosen) and collectively exhaustive (all possible alternatives must be included in the set).

Although PTI's management has been operating on the assumption that they would be solely responsible for the development of this HP-273 synthetic protein throughout this process (Plan A), there are two major decisions in the PTI decision-making problem.

First, PTI has received an offer to sell its HP-273 patent for $4 million now, before undertaking the any costs in the first stage of the approval process (Plan B). The company must decide whether or not to accept the offer.

Second, if PTI succeeds in getting preliminary FDA certification at the end of the first stage, then it could to sell a 50% share of the HP-273 project to other investors for $25 million

(Plan C). The company must decide whether or not to sell a 50% share of the HP-273 project to other investors.

An event node is a point where uncertainty is resolved (a point where the decision maker learns about the occurrence of an event). An event node, sometimes called a "chance node," is shown as a circle. The event set consists of the event branches extending from an event node, each branch representing one of the possible events that may occur at that point. The set of events must be mutually exclusive (if one occurs, the others cannot occur) and collectively exhaustive (all possible events must be included in the set). Each event is assigned a subjective probability; the sum of probabilities for the events in a set must equal one.

The five sources of uncertainty in the PTI decision-making problem are:

1. The cost of the first stage for setting up a pilot plant and testing the synthetic protein on animal subjects is equally likely to be above or below $9 million, has probability 0.25 of being below $7 million, and has probability 0.25 of being above $11 million.

2. It is uncertain whether PTI can successfully get FDA's preliminary certification of safety at the end of this first stage. The probability of successfully getting a preliminary certification of safety at the end of this first stage is 0.3.

3. The cost of taking the synthetic protein though the second-stage extensive medical study has a net present value that is equally likely to be above or below $20 million, has probability 0.25 of being below $16 million, and has probability 0.25 of being above $24 million.

4. Given first-stage certification, if PTI takes the synthetic HP-273 protein through this second stage then the conditional probability of getting full FDA approval for its pharmacological use is 0.6.

5. Given full FDA approval, the returns to PTI from selling its patent would have a present value that is equally likely to be above or below $120 million, has probability 0.25 of being below $90 million, and has probability 0.25 of being above $160 million.

In general, decision nodes and branches represent the controllable factors in a decision problem; event nodes and branches represent uncontrollable factors.

Decision nodes and event nodes are arranged in order of subjective chronology. For example, the position of an event node corresponds to the time when the decision maker learns the outcome of the event (not necessarily when the event occurs).

The third kind of node is a terminal node, representing the final result of a combination of decisions and events. Terminal nodes are the endpoints of a decision tree, shown as the end of a branch on hand-drawn diagrams and as a triangle or vertical line on computer-generated diagrams.

Table 3.1 shows the three kinds of nodes and two kinds of branches used to represent a decision tree.

Table 3.1 Node Symbols and Successors

Type of Node	Written Symbol	Computer Symbol	Node Successor
Decision	square	Square	decision branches
Event	circle	Circle	event branches
Terminal	endpoint	Triangle or vertical line	terminal value

Terminal Values

Each terminal node has an associated terminal value, sometimes called a payoff value, outcome value, or endpoint value. Each terminal value measures the result of a *scenario*: the sequence of decisions and events on a unique path leading from the initial decision node to a specific terminal node.

To determine the terminal value, one approach assigns a cash flow value to each decision branch and event branch and then sum the cash flow values on the branches leading to a terminal node to determine the terminal value. In the PTI problem, there are distinct cash flows associated with many of the decision and event branches. Some problems require a more elaborate value model to determine the terminal values.

Figure 3.3 shows the arrangement of branch names, probabilities, and cash flow values on an unsolved tree.

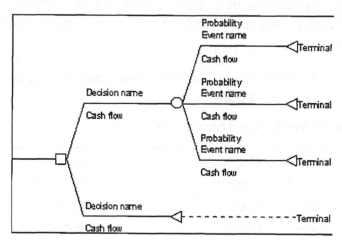

Figure 3.3 Generic decision tree.

PTI Decision Tree

Figure 3.4 shows the unsolved decision tree model for the decision-making problem in PTI, a bioengineering firm.

For example, the +$75 million terminal value on the far right of the diagram is associated with the scenario shown in Table 3.2.

Table 3.2 Terminal Value of a Scenario

Branch type	Branch name	Cash flow
Event	Stage 1 certified	- $9 million
Decision	Sell 50% of share (Plan C)	+$25 million
Event	Stage 2 approved	-$10 million
Event	Getting revenue	+$69 million
	Terminal value	*+75 million*

Planning and Payoff Distribution

A plan specifies an initial choice and any subsequent choices to be made by the decision maker. The subsequent choices usually depend upon events. The specification of a plan must be comprehensive; if the decision maker gives the plan to a colleague, the colleague must know exactly which choice to make at each decision node. Most decision problems have many possible plans, and a goal of the analysis is to determine the optimal plan, taking into account the decision maker's risk attitude.

As illustrated in PTI case study, each plan has an associated payoff distribution, sometimes called a risk profile. The payoff distribution of a particular plan is a probability distribution showing the probability of obtaining each terminal value associated with a particular plan.

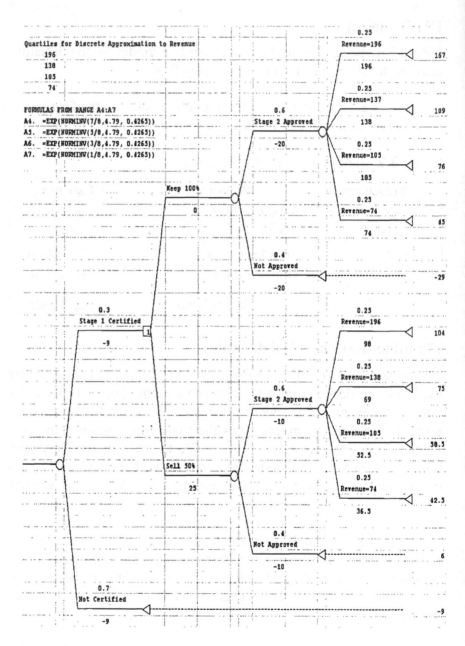

Figure 3.4 Decision tree for PMI case study.

In decision tree models, the payoff distribution can be shown as a list of possible payoff values, x, and the discrete probability of obtaining each value, $P(X=x)$, where X represents the uncertain terminal value. Since a plan specifies a choice at each decision node, the uncertainty about terminal values depends only on the occurrence of events. The probability of obtaining a specific terminal value equals the product of the probabilities on the event branches on the path leading to the terminal node.

The stage 1 cost and the stage 2 cost both have quartiles that satisfy the equal differences property (Q3-Q2=Q2-Q1), so they are Normal random variables.

	Mean	Stdev	Simulation
Stage 1 Cost	9	2.963	NORMINV(RAND(),9, 2.963)
Stage 2 Cost	20	5.926	NORMINV(RAND(),20, 5.926)

Final revenue has quartiles that satisfy the equal ratios property (Q3/Q2=Q2/Q1) and so it is a Lognormal random variable.

	Log-mean	Log-stdev	Simulation
Final Revenue	4.787	0.4262	EXP(NORMINV(RAND(),4.787,0.4262))

The estimated expected profit is higher for Plan A than Plan C, so we can recommend Plan A under the expected value criterion. The expected profit from Plan A is greater than 4, so we can recommend Plan A over the Plan-B alternative of selling HP-273 for $4 million now. To maximize expected profit, the PTI should not sell HP-273 now for less than the expected profit from Plan A, which we estimate to be $9.4042 million.

The probability that the stage 1 cost will be between $8 and $12 million can be calculated by:

=NORMSDIST((12-9)/2.963)-NORMSDIST((8-9)/2.963)
=0.476

and the probability that the stage 1 cost will be more than $12 million can calculated

=1-NORMSDIST((12-9)/2.963)
= 0.156

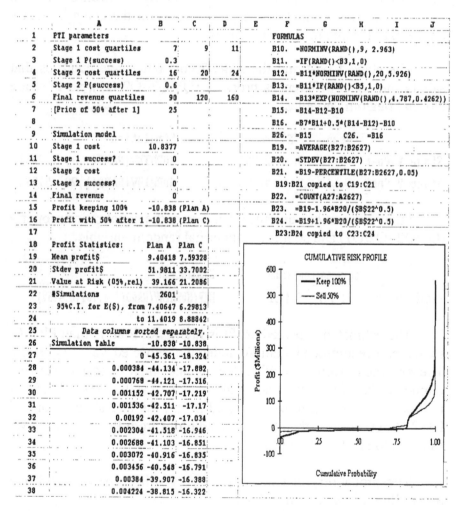

Figure 3.5 Decision tree analysis with spreadsheet simulation.

The cumulative risk profile in Figure 3.5 was made by sorting the simulated profit data in B27:B2627 and plotting it in an XY-chart against the percentile index in A27:A2627. (Because of this sorting, Figure 3.5 shows only the worst profits that were sorted to the top of the data range.) Reading from this chart at the 0.05 and 0.10 cumulative-probability levels we find values at risk of about –$29.7 million and -$21.8 million respectively for Plan A, which are substantially worse than the 5% value at risk found in our analysis of part C (-$13.6 million). However, the expected profit from Plan A ($9.4042 million) is higher than Plan C ($7.5933 million). The decision would be a trade-off between the expected profit and the risk.

3.6 CAPITAL BUDGETING

Capital budgeting is the analysis of long-term projects, such as the case study for a biotech project in Section 3.1. Long-term projects are worthy of special attention because of the fact that they frequently require large initial investments, and because the cash outlay to start such projects often precedes the receipt of profits by a significant period of time. In such cases, we are interested in being able to predict the profitability of the project. We want to be sure that the profits from the project are greater than what we could have received from alternative investments or uses of our money.

This section focuses on how engineers can evaluate long-term projects and determine whether the expected return from the projects is great enough to justify taking the risks that are inherent in long-term investments. Several different approaches to capital budgeting will be discussed in Chapter 6. These are the pay-back method, the net present value method, and the internal rate of return method. The latter two of these methods requires us to acknowledge the implications of the "time value of money." We have indirectly alluded to such a time value of money previously, pointing out that we would prefer to defer tax payments to the future. Such deferment allows us the use of the money in the interim to earn additional profits.

Engineers can use a formalized approach to evaluate the time value of money. Such an approach is necessary because many of the problems we face in capital budgeting cannot be solved easily without a definite mechanical methodology. To give a rather elementary example, suppose that someone offered to buy your product for $250, and that they are willing to pay you either today, or one year from today. You will certainly prefer to receive the $250 today. At the very least, you could put the $250 in a bank and earn interest in the intervening year.

Suppose, however, that the buyer offered you $250 today or $330 in 22 months. Now your decision is much more difficult. How sure are you that the individual will pay you 22 months from now? Perhaps he or she will be bankrupt by then. What could we do with the money if we received it today? Would we put the $250 in some investment that would yield us more than $330 in 22 months from today? These are questions that we have to be able to answer in order to evaluate long-term investment opportunities. But first let's discuss some basic issues of investment analysis.

Investment Opportunities

The first step that must be taken in investment analysis is to generate various investment opportunities. Some firms choose to evaluate projects on an individual basis, accepting any project that can meet certain criteria. Other firms take the approach of comparing all of their alternative opportunities and selecting the best projects from among all of the alternatives.

In either case, the key first step is the development of the investment opportunity. Such opportunities fall into two major classes: new project investments, and replacement or reinvestment in existing projects. New project ideas can come from a variety of sources. They may be the result of research and development activity or exploration. Your firm may have a department solely devoted to new product development. Ideas may come from outside of the firm. Reinvestment is often the result of production managers pointing out that certain equipment needs to be replaced. Such replacement should not be automatic. If a substantial outlay is required, it may be an appropriate time to reevaluate the product or project to determine if the profits being generated are adequate to justify continued additional investment.

Data Generation

The data needed to evaluate an investment opportunity are the expected cash flows related to the investment. Many projects have a large initial cash outflow as we acquire plant and equipment, and incur start-up costs prior to actual production and sale of our new product. In the years that follow, there will be receipt of cash from the sale of our product (or service) and there will be cash expenditures related to the expenses of production and sales. We refer to the difference between each year's cash receipts and cash expenditures as the net cash flow for that year.

You're probably wondering why we have started this discussion with cash flow instead of net income for each year. There are several important reasons. First, net income, even if it were a perfect measure of profitability, doesn't consider the time value of money. For instance, suppose that we have two alternative projects. The first project requires that we purchase a machine for $10,000 in cash. The machine has a ten-year useful life and generates revenue of $1,500 per year. If we depreciate on a straight-line basis for financial reporting, we have a depreciation expense of $1,000 per year. Assume for this simple example that there are no other expenses besides depreciation. Because we have revenues of $1,500 per year and expenses of $1,000 per year, we have a pretax profit of $500 each year.

Suppose that a totally different project requires that we lease a machine for $1,000 a year for ten years, with each lease payment being made at the start of the year. The leased machine produces revenues of $1,500 a year, leaving us with a pretax profit of $500 per year. Are the two alternative projects equal? No, they aren't.

Even though they both provide the same pretax profit, they are not equally as good. One project requires us to spend $10,000 at the beginning. The other project only requires an outlay of $1,000 in the first year. In this second project, we could hold on to $9,000 that had been spent right away in the first project. That $9,000 can be invested and can earn additional profits for the firm.

This issue stems from the generally accepted accounting principle that requires a matching of revenue and expense. Because the machine generates revenue for each of ten years, we must allocate the expense over the ten-year period when we report to our stockholders. The income statement for a firm that matches (that is, uses the accrual basis of accounting) is not based on how much cash it receives or spends in a given year.

There is another problem with the use of net income in this example. Although the pretax income is the same for each of the two projects, the amount of taxes paid by the firm would not be the same for these two projects. The first project provides us with ownership of a machine that has a ten-year useful life, but a 5-year life for tax depreciation under the Modified Accelerated Cost Recovery System (MACRS) discussed in Handouts 9 and 10. The depreciation reported to the IRS doesn't have to equal the depreciation reported to the stockholders. That means more depreciation will be reported to the government than to our stockholders in the early years of the asset's life. The result will be a tax deferment.

Lease payments cannot provide us with tax deductions as great as the MACRS depreciation. Therefore, the project in which we lease instead of purchasing results in higher tax payments in the early years. The higher tax payments leave less money available for investment in profitable opportunities.

Even aside from these two problems, net income is not a very reliable figure on which to base project evaluation. Consider the LIFO inventory. Using LIFO, we intentionally suppress the reported net income in order to lower our taxes. But in determining whether a project is a good one or not, we want to know the true profitability-- not the amount of profits we are willing to admit to when filling out our tax return. Cash is a good measure of underlying profitability. We can assess how much cash we put into a project and how much cash we ultimately get from the project. Just as importantly, we can determine when the cash is spent and when the cash is received to enable investment in other profitable opportunities.

The data needed for investment or project analysis is cash flow information for each of the years of the investment's life. Naturally we cannot be 100 percent certain about how much the project will cost and how much we will eventually receive. There is no perfect solution for the fact that we have to make estimates. However, we must be aware at all times that because our estimates may not be fully correct there is an element of risk. Project analysis must be able to assess whether the expected return can compensate for the risks we are taking.

Our analysis is somewhat simplified if we prepare all of our es- timates on a pretax basis. Taxes can be quite complex and they add extra work on to the analysis. Unfortunately, we would be making a serious error if we didn't perform our analysis on an after-tax basis. Consider the previous example of two alternative projects. The tax payment is not the same for the two alternatives. In order to clearly see which project is better, we must analyze the impact of projects on an after-tax basis.

One final note prior to looking at the methods for assessing al- ternative investments. Our analysis should always attempt to con- sider firm-wide effects of a particular project. Too often we apply tunnel vision to a project. Most of the firm's products bear some re- lationship to the firm's other products. When we introduce a new product, its profitability can't be assessed in a vacuum. For instance, when a toothpaste manufacturer creates a new brand of toothpaste, the hope is that the new brand will gain sales from the competition.

The total market for toothpaste is relatively fixed. Although new brands may increase the total market slightly, most of their sales will come at the expense of other brands already on the market. If the other brands are made by our competitors, great. But it's entirely probable that each of the brands of toothpaste our company already makes will lose some ground to our new brand. When we calculate the profits from the new brand, we should attempt to assess, at least qualitatively if not quantitatively, what the impact on our current brands will be.

On the other hand, consider a shoe manufacturer. If styles switch from loafers (shoelaceless) to wingtips (shoes that need shoelaces), then we can expect our shoelace division to have increased sales. For a sports equipment manufacturer, if a new aluminum baseball bat can expand the baseball bat market, then we can expect sales of baseballs and baseball gloves to increase. New products may have a positive or a negative effect on the rest of the company. In estimating the cash flows of the potential investment, this factor should be considered. The different approaches to capital budgeting will be discussed in detail in Chapter 6.

3.7 SUMMARY

In this chapter we studied families of continuous probability distributions that are commonly used to describe uncertainty about unknown quantities which can take infinitely many possible values.

The Normal probability distributions are generally paremeterized by their mean and standard deviation, and can be applied to unknowns (like sample means) that are the sum of many small terms drawn independently from some distribution. We learned to use Excel's NORMSDIST function to compute the probability of any given interval for a Normal random variable with any mean and standard deviation. The Excel function NORMINV was used to simulate Normal random variables, and plot their inverse cumulative probability curves. After introducing the concept of a probability density curve for a continuous random variable, we showed how to estimate it from an inverse cumulative function like NORMINV.

A Lognormal random variable is one whose natural logarithm is a Normal random variable. Lognormal random variables can be parameterized by their log-mean m and log-standard-deviation s, or by their true mean μ and true standard deviation F. With the first parameterization, they can be simulated by the Excel formula EXP(NORMINV(RAND(),m, s)). Lognormal distributions can be applied to unknown quantities (like fatigue degradation ratios over some period of time) that are the product of many factors, each close to 1, that are drawn independently from some distribution.

Both Normal distribution and Lognormal distribution can be parameterized by the three quartile boundary points. Normal distributions are selected when the quartiles points have equal differences. Lognormal distributions are selected when the quartile points have equal ratios.

REFERENCES

Ahuja, H. N., Dozzi, S. P., Abourizk, S. M. (1994), Project Management, Second Edition, John Wiley & Sons, Inc., New York, NY.

Ang, A. H-S., Tang, W. H. (1984), Probability Concepts in Engineering Planning and Design, Volume II – Decision, Risk, and Reliability," John Wiley & Sons, New York, NY.

AT&T and the Department of the Navy (1993), Design to Reduce Technical Risk, McGraw-Hill Inc., New York, NY

Bell, D, E., Schleifer, A., Jr (1995), Decision Making Under Uncertainty, Singular Pub Group, San Diego, California,

Burlando, Tony (1994), "Chaos and Risk Management," Risk Management, Vol. 41 #4, pages 54-61.

Catalani, M. S., Clerico, G. F. (1996), "Decision Making Structures : Dealing With Uncertainty Within Organizations (Contributions to Management Science)," Springer Verlag, Heidelberg, Germany, January.

Chacko, G. K. (1993), Operations Research/Management Science : Case Studies in Decision Making Under Structured Uncertainty McGraw-Hill, New York, NY.

Chicken, John C. (1994), Managing Risks and Decisions in Major Projects, Chapman & Hall, London, Great Britain.

Cooper, Dale F. (1987), Risk Analysis for Large Projects: Models, Methods, and Cases, Wiley, New York, NY.

Covello, V. T. (1987), Uncertainty in Risk Assessment, Risk Management, and Decision Making (Advances in Risk Analysis, Vol 4), Plenum Pub Corp,

Englehart, Joanne P. (1994), "A Historical Look at Risk Management," Risk Management. Vol.41 #3, pages 65-71.

Esenberg, Robert W. (1992), Risk Management in the Public Sector, Risk Management, Vol. 39 #3, pages 72-78.

Galileo (1638), Dialogues Concerning Two New Sciences, Translated by H. Crew and A. de Salvio, 1991, Prometheus Books, Amherst, New York, NY.

Grose, Vernon L. (1987), Managing Risk: Systematic Loss Prevention for Executives, Prentice-Hall, Englewood Cliffs, New Jersey.

Johnson, R. A. (1994), Miller & Freund's Probability & Statistics for Engineers, Fifth Edition, Prentice Hall, New Jersey.

Klapp, M. G. (1992), "Bargaining With Uncertainty: Decision-Making in Public Health, Technologial Safety, and Environmental Quality," March, Auburn House Pub., Westport, Connecticut.

Kurland, Orim M. (1993), "The New Frontier of Aerospace Risks." Risk Management. Vol. 40 #1, pages 33-39.

Lewis, H.W. (1990), Technological Risk, Norton, New York, NY.

McKim, Robert A. (1992), "Risk Management: Back to Basics." Cost Engineering. Vol. 34 #12, pages 7-12.

Moore, Robert H. (1992), "Ethics and Risk Management." Risk Management. 39 #3, pages 85-92.

Moss, Vicki. (1992), "Aviation & Risk Management." Risk Management. Vol. 39 #7, pages 10-18.

Petroski, Henry (1994), Design Paradigms: Case Histories of Error & Judgment in Engineering, Cambridge University Press, New York, NY.

Raftery, John (1993), Risk Analysis in Project Management, Routledge, Chapman and Hall, London, Great Britain.

Schimrock, H. (1991), "Risk Management at ESA." ESA Bulletin. #67, pages 95-98.

Sells, Bill. (1994), "What Asbestos Taught Me About Managing Risk." Harvard Business Review. 72 #2, pages 76-90.

Shaw, Thomas E. (1990), "An Overview of Risk Management Techniques, Methods and Application." AIAA Space Programs and Technology Conference, Sept. 25-27.

Smith, A. (1992), "The Risk Reduction Plan: A Positive Approach to Risk Management." IEEE Colloquium on Risk Analysis Methods and Tools.

Sprent, Peter (1988), Taking Risks: the Science of Uncertainty, Penguin, New York, NY.

Toft, Brian (1994), Learning From Disasters, Butterworth-Heinemann.

Wang, J. X. and Roush, M. L. (2000), What Every Engineer Should Know About Risk Engineering and Management, Marcel Dekker Inc., New York, NY.

Wideman, R. Max., ed. (1992), Project and Program Risk Management: A Guide to Managing Project Risks and Opportunities, Project Management Institute, Drexel Hill, PA.

4

Correlation of Random Variables and Estimating Confidence

The correlation among engineering variables often puzzles engineers for making decisions under uncertain situations. To balance the technical risk, schedule risk, and cost risk involved in their projects, many decision variables are highly correlated. To assess the relationship between two uncertain engineering variables, we should answer the question about how our beliefs about one variable would be affected by getting information about the other variable. As shown in this chapter, modeling the correlation among engineering variables will help engineers use information effectively to make effective engineering decisions.

4.1 CASE STUDY: A TRANSPORTATION NETWORK

Consider a transportation network between cities A and B consisting of two branches as shown in Figure 4.1.

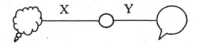

Figure 4.1 A transportation network.

Let X and Y be the travel times over the two respective branches. Because of uncertainties in the traffic and weather conditions, these travel times are modeled as random variables as follows:

Travel Time	Mean	Standard Deviation
X	μ	σ
Y	$c_2\mu$	$D_2\sigma$

Where c_2 and d_2 are constants dependent on the specific network. Furthermore, since X and Y will be subject to common evironmental factors, they are expected to be partially correlated. Suppose these correlations are

$$\rho_{xy} = 0.5$$

The performance function may be defined as

$$g(X) = T_0 - (X + Y)$$

where

$$T_0 = \text{The scheduled total travel time between A and B.}$$

The required logistic cost, W, can be calculated by:

$$W = 100*X + 200*Y \qquad (4.1)$$

Because of the uncertainties with the weather conditions, the exact values of the travel times are unknown quantities. So engineers begin to think about probability distributions for these unknown quantities. Suppose that an logistic engineer assesses the following simple discrete probability distributions for these two unknown quantities:

X	P(X=x)	Y	P(Y=y)
70	0.2	30	0.3
75	0.5	32	0.6
80	0.3	34	0.1

With the Equation (4.1) and the engineer's assessed probability distribution for **X** and Y, can we compute the probability distribution for W? Unfortunately, the answer is No. For example, try to compute the probability that **W** = $13000, which (with the possible values of **X** and **Y** listed above) can occur only if X=70 and Y=30. We known that P(X=70) is 0.2, and P(Y=30) is 0.3, but what is the probability of getting X=70 and Y=30 at the same time? The answer to that question depends on what the engineer believes about the relationship between **X** and Y. If she believes that **X** tends to be low when **Y** is low then the probability of getting both X=70 and Y=30 could be as much as 0.2. But if she believes that **X** will be low only when **Y** is not low, then the probability of getting both X=70 and Y=30 could be 0.

Engineers use the symbol "∩" to denote the intersection of two events, that is, the event that two other events both occur. Unfortunately, the symbol "∩" cannot be used as Excel labels, so we will use the ampersand symbol "&" instead of "∩" here to denote intersections. Thus, the expression "X=70 & Y=30" is used to denote the event that **X** is 70 and **Y** is 30 (so that W=100*70+200*30 = 13000). With this notation, we may say that our problem above is that knowing P(X=70) and P(Y=30) does not tell us enough to compute P(X=70 & Y=30).

A table or function that specifies, for each pair of possible values (x, y), the probability P(X=x & Y=y) is called the joint probability distribution of **X** and Y. To compute the probability distribution, we need to assess the joint probability distribution of **X** and Y.

When **X** and **Y** are independent random variables, then their joint probability distribution can be computed by the simple formula

$$P(X=x \ \& \ Y=y) = P(X=x) * P(Y=y).$$

So for this example, if **X** and **Y** are independent, then P(X=70 & Y=30) is 0.2*0.3 = 0.06. With this multiplicative formula, specifying the separate distributions of independent random variables implicitly gives us their joint distribution. That is why, in Chapter 3, we did not worry about joint probability distributions of the multiple unknown quantities that arose in the "PTI Case Study," because we were assuming that these unknown quantities were all independent of each other.

Because of uncertainties in the traffic and weather conditions, the travel times, X and Y, are correlated with each other. Thus, there is good reason to suppose that our engineer might not think of **X** and **Y** as independent random variables. The discrete distributions shown above allow nine possible pairs of values for **X** and **Y** together

{(70,30), (70,32), (70,34), (75,30), (75,32), (75,34), (80,30), (80,32), (80,34)}

and we may ask our engineer to subjectively assess a probability of each pair. She may have some difficulty assessing these nine probabilities (which must sum to 1 because exactly one pair will occur), but let us suppose that her answers are as summarized in the joint probability table shown in the range A3:D6 of Figure 4.2.

	A	B	C	D	E	F
1	JointProbys	P(X=x&Y=y)				
2		y=				
3	x= \	30	32	34		P(X=x)
4	70	0.1	0.1	0		0.2
5	75	0.2	0.3	0		0.5
6	80	0	0.2	0.1		0.3
7		P(Y=y)				sum
8		0.3	0.6	0.1		1
9						
10	X	Y				
11	75.5	31.6	Expected Value			
12	3.5	1.2	Standard deviation			
13						
14	FORMULAS FROM RANGE A1:F12					
15	F4. =SUM(B4:D4)					
16	F4 copied to F4:F6.					
17	B8. =SUM(B4:B6)					
18	B8 copied to B8:D8.					
19	F8. =SUM(B4:D6)					
20	A11. =SUMPRODUCT(A4:A6,F4:F6)					
21	B11. =SUMPRODUCT(B3:D3,B8:D8)					
22	A12. = SUMPRODUCT((A4:A6-A11)^2,F4:F6)^0.5					
23	B12. = SUMPRODUCT((B3:D3-B11)^2,B8:D8)^0.5					

Figure 4.2 Joint probability distribution of two random variables.

The separate probability distributions for **X** and **Y** can be computed from this joint probability distributions by summing across each row (to get the probability distribution of X) or down each column (to get the probability distribution of Y). These computations are performed in cells F4:F6 and B8:D8 of Figure 4.2, where they yield the probability distributions for **X** and **Y** that we described above. (So this joint distribution is consistent with the earlier assessment of the separate probability distributions of **X** and Y.) The probability distribution of a single random variable is sometimes called its marginal probability distribution when it is computed from a joint distribution in this way, so called simply because it is often displayed next to the margin of the joint probability table, as in this spreadsheet.

In the joint probability distribution table, as shown in Figure 4.2, we can see more probability in the top-left and bottom-right corners, where both unknowns are greater than the expected value or both are less than the expected value, than in the top-right and bottom-left corners, where one unknown is greater than expected and the other is less than expected. Thus, the joint probability table in Figure 4.2 clearly exhibits a kind of positive relationship between X and Y, as stock prices that tend to go up or down together. To provide a quantitative measure of such relationships or tendencies of random variables to co-vary, mathematicians have defined two concepts called covariance and correlation.

4.2 COVARIANCE AND CORRELATION

Before introducing these two statistical concepts useful to an in-depth understanding of engineering decision-making problems, it will be helpful to switch to a different way of tabulating joint probability distributions, because the two-dimensional table shown in Figure 4.2 is not so convenient for computations in spreadsheets. A more convenient (if less intuitive) way of displaying a joint probability distribution is to list each possible combination of values of X and Y in a separate row, with all the joint probabilities listed down one column. Such a one-dimensional joint probability table is shown in the range A4:C13 of Figure 4.3. The nine possible pairs of values of X and Y are listed in rows 5 through 13, with X values in A5:A13 and Y values in B5:B13, and the probabilities for these (X,Y) pairs are listed in cells C5:C13. You should verify that the joint probabilities shown in cells C5:C13 of Figure 4.3 are really the same as those shown in cells B4:D6 of Figure 4.2.

In general, the covariance of any two random variables **X** and **Y** is defined to be the expected value of the product of their deviations from their respective means. That is,

$$\text{Covar}(X,Y) = E((X-\mu_x)*(Y-\mu_y)).$$

For our example, the mean μ_x is computed in cell A16 of Figure 4.3 by the formula

=SUMPRODUCT(A5:A13,$C$5:$C$13)

and the mean μ is computed in cell B16 by copying from cell A16. Then the deviations from the y means $(X-\mu_x)*(Y-\mu_y)$ are computed for the nine possible pairs of (X,Y) values by entering the formula

=(A5-$A$16)*(B5-$B$16)

into cell E5 of Figure 4.3, and then copying E5 to E5:E13. Now to compute the expected product of deviations from the means, we multiply each possible product of deviations by its corresponding probability and sum. So the covariance can be computed by the formula

=SUMPRODUCT(E5:E13,$C$5:$C$13)

in cell E16 of Figure 4.3.

How should we interpret this covariance? Recall first that the product of two numbers is negative only when one is negative and one is positive; the product of two negative numbers is positive. So the product of deviations from means is positive in the cases the two random variables deviate from their respective means in the same direction (that is, when both are greater than their means or both are less than their means). The products of deviations from means is negative only when the two random variables are deviating from their respective means in opposite directions (one greater than its mean while the other is less). So the covariance is a positive number here because there is more probability in the cases where the two random variables are varying from their means in the same direction, and there is less probability in the cases where the two random variables are varying from their means in opposite directions. In this sense, the covariance is a measure of the tendency of two random variables to co-vary.

Notice also that the covariance of a random variable with itself is the random variables variance, which is the square of the standard deviation. That is,

$$Covar(X,X) = E((X-\mu_x)*(X-\mu_x)) = E((X-\mu_x)^2) = Var(X) = Stdev(X)^2.$$

Recall that the variance of **X** was defined as the expected value of the squared deviation of **X** from its own expected value, and the standard deviation was defined as the square root of the variance. Here "Var" stands for "variance."

The actual numerical value of the covariance (2.2 in this example) is hard to interpret because (like the variance) it has a very strange unit of measurement. When our random variables **X** and **Y** are measured in volts, their means and standard deviations are also measured in minutes. But the products of deviations involve multiplying minutes amounts by other minutes amounts, and so the covariance is measured in minutes times minutes or minutes-squared.

But when we divide the covariance of any two random variable by their standard deviations, we get a unit-free number called the correlation coefficient, which will be easier to interpret. That is, the correlation of any two random variables **X** and **Y** is defined by the formula

$$Correl(X,Y) = Covar(X,Y)/(Stdev(X)*Stdev(Y))$$

For our example, the correlation of **X** and **Y** is computed by the formula =E17/(A17*B17) in cell F17 in Figure 4.3.

	1	2	3	4	5	6	7	8	9	10	11
1							Coefficients				
2							of X	of Y			
3		JointProbys					100	200			
4	x	y	P(X=x&Y=y)		ProdDevsFromMean				W = Output value		
5	70	30	0.1		8.8				13000		
6	70	32	0.1		-2.2				13400		
7	70	34	0		-13.2				13800		
8	75	30	0.2		0.8				13500		
9	75	32	0.3		-0.2				13900		
10	75	34	0		-1.2				14300		
11	80	30	0		-7.2				14000		
12	80	32	0.2		1.8				14400		
13	80	34	0.1	sum	10.8				14800		
14					1						
15	X	Y		Covariance	Correlation				W		=100*X+200*Y
16	75.5	31.6 E			2.2	0.5238			13870 E		13870
17	3.5	1.2 Stdev							517.78 Stdev		517.78
18									268100 Var		268100
19				CorrelArray							
20					1	0.5238					
21				0.5238	1						
22											

23	FORMULAS	
24	E5. =(A5-$A$16)*(B5-$B$16)	E5 copied to E5:E13.
25	I5. =SUMPRODUCT($G$3:$H$3,A5:B5)	I5 copied to I5:I13.
26	D14. =SUM(C5:C13)	
27	A16. =SUMPRODUCT(A5:A13,$C$5:$C$13)	A16 copied to B16,E16,I16.
28	A17. =SUMPRODUCT((B6:B14-B17)^2,D6:D14)^0.5	A17 copied to B17,I17.
29	F16. =E16/(A17*B17)	
30	I16. = =SUMPRODUCT(I5:I13,$C$5:$C$13)	
31	I17. =SUMPRODUCT((I5:I13-I16)^2,C5:C13)^0.5	
32	I18. =I17^2	
33	K16. =SUMPRODUCT(G3:H3,A16:B16)	
34	K17. = SUMPRODUCT(G3^2*A17^2+H3^2*B17^2+2*E20*G3*H3*A17*B17)^0.5	
35	K18. = K17^2	

Figure 4.3 Correlation of two discrete random variables and linear-system analysis.

To interpret the correlation of two random variables, you should know several basic facts. The correlation of any two random variables is always a number between -1 and 1. If **X** and **Y** are independent random variables then their correlation is 0. On the other hand, an extreme correlation of 1 or -1 occurs only when X depends on Y by a simple linear formula of the form $X = c*Y + d$ (where c and d are nonrandom numbers). Such a linear relation yields correlation 1 when the constant c is positive, so that X increases linearly as Y increases; but it yields correlation -1 when the constant c is negative, so that X decreases linearly as Y increases.

So when we find that X and Y have a correlation coefficient equal to 0.524 in this example, we may infer the X and Y have a relationship that is somewhere about mid-way between independence (correlation = 0) and a perfect positively-sloped linear relationship (correlation = 1). A higher correlation would imply a relationship that looks more like a perfect linear relationship, while a smaller positive correlation would imply a relationship that looks more like independence.

4.3 LINEAR FUNCTIONS OF SEVERAL RANDOM VARIABLES

We introduced this chapter with an engineer who designs a linear control system. The system output W will depend on X and Y by the linear formula $W = 100*X + 200*Y$. In Figure 4.2, the possible output values corresponding to each possible (X,Y) pair are shown in column I. To compute the value of W that corresponds to the (X,Y) values in cells A5:B5, cell I5 contains the formula

=SUMPRODUCT($G$3:$H$3,A5:B5),

and this formula from cell I5 has been copied to the range I5:I13. Then E(W) and Stdev(W) can be computed by the formulas

= SUMPRODUCT(I5:I13,$C5:C13)

=STDEVPR(I5:I13,$C5:C13)

in cells I16 and I17 respectively. The variance Var(W), as the square of the standard deviation, is computed in cell I18 of Figure 4.3 by the formula

$$=I17^2$$

But the expected value and variance of **W** can also be calculated another way, by using some general formulas about the means and variances of linear functions of random variables. the expected value of **W** can be computed by the simple linear formula E(W) = 100*E(X)+200*E(Y). With the coefficients 100 and 200 listed in G3:H3 of Figure 4.3, and with the expected values E(X) and E(Y) listed in A16:B16, this linear formula for E(W) can be computed in cell K16 by the formula

$$=SUMPRODUCT(G3:H3,A16:B16)$$

To check, notice cell K16 and I16 are equal in Figure 4.3.

For the case study in Section 4.1, we have 2 choices for X and 2 choices for Y, and so we have a matrix of 2*2 correlations appearing in this formula. To do computations in a spreadsheet, we will regularly list all these correlations together in a correlation array.

The correlation array for our example in Figure 4.3 is shown in the range D20:E21. With **X** as our first control value and **Y** as our second control value, the correlation array is

Correl(X,X) = 1;

Correl(X,Y) = 0.524;

Correl(Y,X) = 0.524;

Correl(Y,Y) = 1.

In any correlation array, the elements on the diagonal from top-left to bottom-right are always ones, because the correlation of any random variable with itself is 1. Below this diagonal and to its right, the elements of the correlation array are always symmetric, because Correl(X,Y) equals Correl(Y,X).

To calculate the standard deviation, we need to multiply each number in the correlation array by a corresponding product of coefficients, $\alpha_x*\alpha_x$, $\alpha_x*\alpha_y$, $\alpha_y*\alpha_x$, or $\alpha_y*\alpha_y$, and by a corresponding product of standard deviations Std(X)*Std(X), Std(X)*Std(Y), Std(Y)*Std(X), or Std(Y)*Std(Y). The four products of the share coefficients α_x =100 and α_y =200 are

$$100*100 = 10,000;$$
$$100*200 = 20,000;$$
$$200*100 = 20,000;$$
$$200*200 = 40,000.$$

as shown in cells G20:H21 in Figure 4.3. The four products of the standard deviations Std(X) = 3.5 and Std(Y) = 1.2 are

$$3.5*3.5 = 12.25;$$
$$3.5*1.2 = 4.2;$$
$$1.2*3.5 = 4.2;$$
$$1.2*1.2 = 1.44.$$

as shown in cells A20:B21 of Figure 4.3. It would not be difficult to compute all these products by entering multiplication formulas into each cell.

An array formula in Excel is a formula that is shared by a whole range of cells at once. To enter an array formula, first select the desired range (say, A20:B21), then type the desired formula, and finish with the special keystroke combination [Ctrl]-[Shift]-[Enter]. (That is, while holding down the [Ctrl] and [Shift] keys, press [Enter].) Now if you look at the formula for any cell in the range that you selected, you should see the array formula that you entered, within a pair of special braces { and }, which were added by Excel to indicate that this is an array formula. (Typing the braces yourself will not work; you must use [Ctrl]-[Shift]-[Enter]. From any selected cell that has an array formula, you can identify the whole range of cells that share the formula by the special keystroke [Ctrl]-/.)

The standard deviation can now be computed as the square root (that is, the power ^0.5) of the variance. So the formula for Std(W) in cell K17 of Figure 4.3 is

= SUMPRODUCT(G3^2*A17^2+H3^2*B17^2+2*E20*G3*H3*A17*B17)^0.5

4.4 MULTIVARIATE NORMAL RANDOM VARIABLES

The Multivariate-Normal distributions are a family of joint probability distributions for collections of two or more random variables. To specify a Multivariate-Normal distribution some collection of random variables X_1, ..., X_n, we must specify the mean $E(X_i)$ and standard deviation $Stdev(X_i)$ for each of these random variables, and the correlation $Correl(X_i, X_i)$ for each pair of these random variables.

Suppose that $(X_1,...,X_n)$ are Multivariate-Normal random variables; that is, their joint probability distribution is in this Multivariate-Normal family. Then each of these random variables X_i is (by itself) a Normal random variable. But more importantly, any linear function of these random variables will also be a Normal random variable. That is, given any nonrandom numbers α_0, α_1, ..., α_n, the linear function

$$A = \alpha_0 + \alpha_1*X_1 + ... + \alpha_n*X_n$$

will also have a Normal probability distribution.

This fact, that every linear function of Multivariate-Normals is also Normal, makes the Multivariate-Normal joint distribution very convenient for financial analysis. We have seen how the mean and standard deviation of a linear-control system's value can be computed from the means, standard deviations, and correlations of the various control values. So if we can assume that these control values have a Multivariate-Normal joint distribution, then our linear control system will have the Normal distribution with this mean and standard deviation, and so we can easily compute its probability of being greater or less than any given number.

Any collection of independent Normal random variables is itself a collection of Multivariate-Normal random variables (with correlations that are zero for all distinct pairs). So any linear function of independent Normal random variables is also Normal. When we compute several different linear functions of the same independent Normal random variables, the results of these linear functions are Multivariate-Normal random variables that have means, standard deviations, and correlations that can be computed according to Section 4.3.

This method of creating Multivariate-Normal random variables is illustrated in Figure 4.4. The formula =NORMINV(RAND(),0,1) copied into cells A5:A8 first gives us four independent Normal random variables, each with mean 0 and standard deviation 1. Entering the formula

=C3+SUMPRODUCT(C5:C8,$A$5:$A$8)

into cell C11, and copying C11 to C11:E11, we get a collection of three non-independent Multivariate-Normal random variables in cells C11:E11, each generated by a different linear function of the independent Normals in A5:A8.

The expected values for the three Multivariate-Normals in cells C11:E11 are just the constant terms from C3:E3, because the underlying independent Normals in A5:A8 all have expected value 0. The standard deviations for the three Multivariate-Normals in C11:E11 are computed in cells C14:E14 (simplified by the fact that each of the independent Normals in A5:A8 has standard deviation 1).

	A	B	C	D	E
	MAKING MULTIVARIATE-NORMAL RANDOM VARIABLES				
			Constants		
			10	20	30
	Independent (0,1)-Normals	Linear coefficients			
	0.362623496		5	7	0
	1.564922059		3	4	-4
	0.611159976		6	0	9
	1.523644642		-2	5	2
		Multivariate-Normal random variables			
			X	Y	Z
			17.12755	36.41628	32.28804
		Expected values	10	20	30
		Standard deviations	8.602325	9.486833	10.04988
		Correlation array	X	Y	Z
		X	1	0.453382	0.439549
		Y	0.453382	1	-0.06293
		Z	0.439549	-0.06293	1

Any linear function of Multivariate-Normals is also Normal.

	Constant	1000		
	Coefficients	100	300	200
W = LinearFunction(X,Y,Z)		20095.25		
	E(W)	14000		
	Stdev(W)	3987.43		

FORMULAS
A5. =NORMINV(RAND(),0,1)
A5 copied to A5:A8
C11. =C3+SUMPRODUCT(C5:C8,$A$5:$A$8)
C13. =C3
C14. =SUMPRODUCT(C5:C8,C5:C8)^0.5
C11:C14 copied to C11:E11
D17. =SUMPRODUCT(C5:C8,D5:D8)/(C14*D14)   C18. =D17
E17. =SUMPRODUCT(C5:C8,E5:E8)/(C14*E14)   C19. =E17
E18. =SUMPRODUCT(D5:D8,E5:E8)/(D14*E14)   D19. =E18
C24. =C22+SUMPRODUCT(C23:E23,C11:E11)
C25. =C22+SUMPRODUCT(C23:E23,C13:E13)
C26. =SUMPRODUCT(PRODS(C23:E23),PRODS(C14:E14),C17:E19)^0.5

Figure 4.4 Creating Multivariate-Normal random variables.

The correlations among the three Multivariate-Normals in C11:E11 are computed in cells D17, E17, and E18. Notice that the random variables in cells C11 and D11 have a positive correlation (0.453) mainly because C11 and D11 both depend with positive co-efficients on the independent Normals in cells A5 and A6, so that any change in A5 or A6 would push C11 and D11 in the same direction. But the coefficients in cells C8 and D8 have different signs, and so any change in cell A8 would push C11 and D11 in opposite directions, which decreases the correlation of C11 and D11. The correlation between cells D11 and E11 is negative (-0.0629) mainly because the random cell A6 has opposite effects on D11 and E11, due to the coefficients 4 and -4 in cells D6 and E6.

4.5 ESTIMATING CORRELATIONS FROM DATA

Excel provides a statistical function CORREL for estimating the correlation among random variables that are drawn from some joint distribution which we do not know but from which we have observed repeated independent draws. To illustrate the use of this statistical function, Figure 4.5 shows travel times for six major transportation that are related a large transportation network for the last ten years. Then the annual increase ratio for the number of travel times is then computed in range B17:G26, by entering the formula =B4/B3 in cell B17 and then copying cell B17 to B17:G26.

Our goal is to use this increase-ratio data for the last ten years to build simple forecasting models that could help to make predictions about the travel times for the coming years. The basic assumption of our analysis will be that the annual increase ratios of these travel times are jointly drawn each year out of some joint probability distribution, which we can try to estimate from the past data. Then we can use this probability distribution to make a simulation model of future forecasting.

Travel times

	Branch X_1	Branch X_2	Branch X_3	Branch X_4	Branch X_5	Branch X_6
Year 1	65	47	38	61	84	103
Year 2	79	61	37	64	89	119
Year 3	85	73	39	74	93	128
Year 4	78	60	40	72	88	120
Year 5	107	89	47	95	119	160
Year 6	108	86	46	89	118	160
Year 7	124	104	57	114	133	187
Year 8	156	120	71	147	179	236
Year 9	195	140	74	146	228	287
Year 10	181	134	72	127	206	266
Year 11	216	175	87	152	251	323

Annual increase ratios

	Branch X_1	Branch X_2	Branch X_3	Branch X_4	Branch X_5	Branch X_6
Year 1	1.22	1.28	0.98	1.05	1.06	1.15
Year 2	1.09	1.20	1.04	1.14	1.05	1.08
Year 3	0.92	0.83	1.02	0.98	0.95	0.93
Year 4	1.37	1.48	1.18	1.32	1.35	1.34
Year 5	1.01	0.96	0.99	0.94	0.99	1.00
Year 6	1.15	1.22	1.22	1.28	1.13	1.16
Year 7	1.26	1.15	1.26	1.29	1.35	1.26
Year 8	1.25	1.17	1.04	1.00	1.27	1.21
Year 9	0.93	0.96	0.97	0.87	0.91	0.93
Year 10	1.19	1.31	1.22	1.19	1.22	1.21

FORMULAS

B17. =B4/B3
  B17 copied to B17:G26.

Figure 4.5 Travel times for six transportation branches.

We have already seen how the unknown means and standard deviations for each random variable can be estimated from our statistical data by the sample average and the sample standard deviation, which can be computed in Excel by the functions AVERAGE and STDEV. In Figure 4.6, for example, the annual increase-ratio data for Branch 1 is exhibited in the range B2:B11. So cell B13 estimates the mean annual growth ratio for Branch 1 by the formula =AVERAGE(B2:B11), and cell B14 estimates the standard deviation of Branch 1's annual growth ratio. Cells B18:B30 remind us of how the STDEV function computes this estimated standard deviation: by computing the squared deviation of each data point from the sample mean, then computing an adjusted average of these squared deviations ("adjusted" in that we divide by $n-1$ instead of n), and finally returning the square root of this adjusted-average of the squared deviations.

The annual increase-ratio data for Branch 2 is exhibited in the range C2:C11 of Figure 4.5. So copying cells B13 and B14 to C13 and C14 similarly yields our statistical estimates of the mean and standard deviation of the annual increase ratio for Branch 2.

Now to estimate the correlation between the annual increase ratios of Branches 1 and 2, the formula

=CORREL(B2:B11,C2:C11)

has been entered into cell B15 in Figure 4.5. The E column in Figure 4.5 shows in more detail how Excel's CORREL function computes this estimated correlation. First, the product of the two random variables' deviations from their respective means is computed, which is done in the spreadsheet of Figure 4.5 by entering the formula

=(B2-$B$13)*(C2-$C$13)

into cell E2 and then copying cell E2 to E2:E11. Next, the adjusted average (dividing by $n-1$ instead of n, where n is the number of observations that we have of each random variable) of these products of deviations is computed, as shown in cell E13 of Figure 4.4 by the formula

$$=SUM(E2:E11)/(10-1)$$

This adjusted-average of the observed products of deviations is a good estimate of the covariance of these two random variables (which has been defined as the expected product of their deviations from their respective means). But the correlation is the covariance divided by both of the standard deviations. So we can compute the estimated correlation from the estimated covariance in cell E15 by the formula

$$=E13/(B13*C13)$$

Notice that this formula in cell E15 returns the same value as the CORREL function in cell B15 of Figure 4.4.

By the law of large numbers, if we get a very large sample of paired random variables (like the increase ratios of Failure Modes 1 and 2 here) that are drawn together from some fixed joint distribution, with each pair drawn independently of all other pairs, then the sample correlation computed by CORREL is very likely to be very close to the true correlation of the joint probability distribution from which we have sampled.

4.6 ACCURACY OF SAMPLE ESTIMATES

When we estimate an expected value by computing the average of sample data, we need to know something about how accurate this estimate is likely to be. Of course the average of several random variables is itself a random variable that has its own probability distribution. As we have discussed, for any two numbers μ and α such that $\alpha>0$, there is precisely-defined Normal probability distribution that has mean (or expected value) μ and standard deviation α.

	A	B	C	D	E	F	G	H
1	Annual Increase Ratio	Branche X₁	Branch X₂		Products of deviations from means			
2	Year 1 to 2	1.22	1.28		0.0098			
3	Year 2 to 3	1.09	1.20		-0.0022			
4	Year 3 to 4	0.92	0.83		0.0718			
5	Year 4 to 5	1.37	1.48		0.0768			
6	Year 5 to 6	1.01	0.96		0.0260			
7	Year 6 to 7	1.15	1.22		0.0005			
8	Year 7 to 8	1.26	1.15		-0.0004			
9	Year 8 to 9	1.25	1.17		0.0019			
10	Year 9 to 10	0.93	0.96		0.0421			
11	Year 10 to 1	1.19	1.31		0.0087			
12								
13	Means	1.1375	1.1548		0.0261 Sum/(10-1) [Covariance]			
14	StDevns	0.1520	0.1922					
15	Correl	0.8937			0.8937 Correlation			
16								
17		Squared deviations from means						
18		0.0060	0.0157					
19		0.0027	0.0019					
20		0.0495	0.1042					
21		0.0561	0.1052					
22		0.0175	0.0386					
23		0.0001	0.0037					
24		0.0148	0.0000					
25		0.0134	0.0003					
26		0.0445	0.0398					
27		0.0033	0.0231					
28								
29	Sum/(10-1)	0.0231	0.0370					
30	SquareRoot	0.1520	0.1922					
31								
32	FORMULAS FROM RANGE A1:E31							
33	B13. =AVERAGE(B2:B11)							
34	C13. =AVERAGE(C2:C11)							
35	B15. =CORREL(B2:B11,C2:C11)							
36	B18. =(B2-B$13)^2							
37	B18 copied to B18:C27							
38	B29. =SUM(B18:B27)/(10-1)							
39	C29. =SUM(C18:C27)/(10-1)							
40	B30. =B29^0.5							
41	C30. =C29^0.5							
42	E2. =(B2-$B$13)*(C2-$C$13)							
43	E2 copied to E2:E11							
44	E13. =SUM(E2:E11)/(10-1)							
45	E15. =E13/(B14*C14)							

Figure 4.6 Computing the correlation of travel times X_1 and X_2.

In an Excel spreadsheet, we can make a random variable that has a Normal probability distribution with mean μ and standard deviation α by the formula

$$=NORMINV(RAND(),\mu,\alpha)$$

Once you know how to make a spreadsheet cell that simulates a given probability distribution, you can learn anything that anybody might want to know about this distribution by simulating it many times in a spreadsheet. So if you want to know what a "Normal distribution with mean 100 and standard deviation 20" is like, you should simply copy the formula $=NORMINV(RAND(),100,20)$ into a large range of cells and watch how the values of these cells jump around whenever you press the recalc key [F9].

As discussed in Chapter 3, a random variable $\mathbf{X}$ has a Normal probability distribution with mean μ and standard deviation α (where μ and α are given numbers such that $\alpha>0$), then

$$P(X<\mu) = 0.5 = P(X>\mu),$$

$$P(\mu-\alpha<X<\mu+\alpha) = 0.683$$

$$P(\mu-1.96*\alpha<X<\mu+1.96*\alpha) = 0.95$$

$$P(\mu-3*\alpha<X<\mu+3*\alpha) = 0.997$$

That is, a Normal random variable is equally likely to be above or below its mean, it has probability 0.683 of being less than one standard deviation away from its mean, it has probability 0.997 (almost sure) of being less than 3 standard deviations of its mean. And for constructing 95% confidence intervals, we will use the fact that a Normal random variable has probability 0.95 of being within 1.96 standard deviations from its mean.

The central limit theorem tells us that Normal distributions can be used to predict the behavior of sample averages.

Consider the average of n random variables that are drawn in-dependently from a probability distribution with expected value μ and standard deviation α. This average, as a random variable, has expected value μ, has standard deviation $\alpha/(n^{\wedge}0.5)$, and has a probability distribution that is approximately Normal.

This central limit theorem is the reason why, of all the formulas that people could devise for measuring the center and the spread of probability distributions, the expected value and the standard deviation have been the most useful for statistics. Other probability distributions that have the same expected value 3 and standard deviation 1.14 could be quite different in other respects, but the central limit theorem tells us that an average of 30 independent samples from any such distribution would behave almost the same. (For example, try the probability distribution in which the possible values are 2, 3, and 7, with respective probabilities $P(2)=0.260$, $P(3)=0.675$, and $P(7)=0.065$.)

Now suppose that we did not know the expected value of K, but we did know that its standard deviation was $F=1.14$, and we knew how to simulate K. Then we could look at any average of n independently simulated values and we could assign 95% probability to the event that our sample average does not differ from the true expected value by more than $1.96*\alpha/(n^{\wedge}0.5)$. That is, if we let Y_n denote the average of our n simulated values, then the interval from $Y_n -1.96*\alpha'(n^{\wedge}0.5)$ to $Y_n +1.96*\alpha/(n^{\wedge}0.5)$ would include the true $E(K)$ with probability 0.95. This interval is called a 95% confidence interval. With $n=30$ and $\alpha=1.14$, the radius r (that is, the distance from the center to either end) of this 95% confidence interval would be

$$r = 1.96*\alpha/(n^{\wedge}0.5) = 1.96*1.14/(30^{\wedge}0.5) = 0.408.$$

If we wanted the radius of our 95% confidence interval around the sample mean to be less than some number r, then we would need to increase the size of our sample so that

$$1.96*\alpha/(n^{0.5}) < r, \text{ and so } n > (1.96*\alpha/r)^2$$

For example, to make the radius of our 95% confidence interval smaller than 0.05, the sample size n must be

$$n > (1.96*\alpha/r)^2 = (1.96*1.14/0.05)^2 = 1997.$$

Now consider the case where we know how to simulate an unknown quantity but we do not know how to calculate its expected value or its standard deviation. In this case, where our confidence-interval formula calls for the unknown probabilistic standard deviation F, we must replace it by the sample standard deviation that we compute from our simulation data. If the average of n independent simulations is X and the sample standard deviation is S, then our estimated 95% confidence interval for the true expected value is from $X-1.96*S/(n^{0.5})$ to $X+1.96*S/(n^{0.5})$, where the quantity $S/(n^{0.5})$ is our estimated standard deviation of the sample average.

A corollary of the central limit theorem can also be used to tell us something about the accuracy of our statistical estimates of points on the (inverse) cumulative probability curve. Suppose that **X** is a random variable with a probability distribution that we know how to simulates, but we do not know how to directly compute its cumulative probability curve. For any number y, let $Q_n(y)$ denote the percentage of our sample that is less than y, when we get a sample of n independent values drawn from the probability distribution of X. Then we should use $Q_n(y)$ as our estimate of P(X<y). But how good is this estimate? When n is large, $Q_n(y)$ is a random variable with an approximately Normal distribution, its expected value is P(X<y), and its standard deviation is $(P(X<y)*(1-P(X<y))/n)^{0.5}$. But this standard deviation is always less than $0.5/(n^{0.5})$. Of course $1.96*0.5/(n^{0.5})$ is only slightly less than $1/n^{0.5}$. So around any

point $(Q_n(y),y)$ in an estimated inverse cumulative probability curve like, we could put a horizontal confidence interval over the cumulative probabilities from $Q_n(y)-1/n^{0.5}$ to $Q_n(y)+1/n^{0.5}$, and this interval would have a probability greater than 95% of including the true cumulative probability at y. When n is 400, for example, the radius of this cumulative-probability interval around $Q_n(y)$ is $1/n^{0.5} = 1/20 = 0.05$. If we wanted to reduce the radius of this 95%- confidence interval below 0.02, then we would increase the sample size to $1/0.02^2 = 2500$.

4.7 CUSTOMER-FOCUSED ENGINEERING DECISION MAKING

Engineering activities exist to serve customers. It's that simple. All else is peripheral to that simple fact. Without customers, engineering activities cannot last. Although well-treated customers are loyal, today customers have choices. In a global business environment, virtually every engineering organization has, or will have competitors. It is insufficient to simply meet the customer's needs once. It must happen every day. Thus, exceeding the customer's expectations every day, in every way, must become a way of life.

What is Responsiveness?

Responsiveness is the ability of an engineering organization to respond quickly to meet the needs of a customer. A synonym for responsiveness is agility. Successful engineering organizations today are able to respond quickly to meet customer's needs for engineering design, production development, and product service. Customer's needs vary. In some cases it may be a new material, subassembly or product. In others, it may be an extremely short turnaround time between when the customer places an order and when he receives the product. This is called *delivery responsiveness*.

   Delivery responsiveness is really a specialized case of responsiveness. It is, most simply, the time it takes from the time the customer places the order until the customer receives the product or service. Frequently, delivery responsiveness is measured with the average.

What is Organizational Reliability?

Organizational reliability is the ability of an organization to meet the commitments made to customers. Commitments to customers should not be made lightly. When a commitment is made, it should be the highest priority of an engineering organization and should always be met. In today's competitive business environment customers have a right to and do expect absolute reliability.

Delivery reliability is a specialized case of reliability. It is of critical importance to you if your customers use JIT inventory management. If your product gets to your customer to early, it will be in the way; if it gets there too late, your late delivery may have shut down the customer's plant. If your deliveries are not reliable, your lack of reliability will force your customer to warehouse a safety stock of your product, thus increasing your customer's costs in both warehousing space and inventory carrying costs. This will make you less competitive from your customer's viewpoint. Delivery reliability is often measured using the standard deviation.

Responsiveness and Reliability from the Customer's Perspective

When engineering organizations look at themselves, they frequently look at themselves much differently than does the customer. For example, when engineering organizations look at responsiveness, most frequently they use measurements internal to them. Two examples will make this clear. 1) Many engineering organizations' sole measure of responsiveness to the customer is to measure their elapsed time against the time interval they promised the customer. They begin counting when the promised interval is completed. Thus, if they promised the customer 10 day delivery, and they actually deliver in 12 days, they are over by 2 days. They think this is pretty good. The customer, however, counts all 12 days, you can be sure. 2) Another engineering measure of responsiveness is from the time when they begin production to the time when the order is shipped. No matter that the order didn't begin production for 7 days. No matter that it takes an additional three days for delivery after production. They tell the customer that they can produce the order in 10 days. It does. They believe they did well. The customer can't figure out how 10 days turned into 20.

If it is important to the customer to have product delivered to them in an extremely short period of time, it is important to do that. If we don't, at some point, an aggressive competitor will. And our aggressive competitor will look at delivery from the customer's perspective.

Knowing What is Important to Customers

How can a engineering organization know what is important to their customers? They aggressively put themselves into the customer's shoes and try to think like the customer and anticipate the customer's needs. They find ways to ask the customer what is going well, what they can do to better meet the customer's needs. They get to know the customer very well and to try to under stand the customer's problems at least as well as the customer knows them.

Focus groups, periodic surveys, and individual meetings with the customer are important venues for finding out how to better please the customer. Customer measures themselves are important enough to merit a special category in a balanced scorecard.

4.8 SUMMARY

This chapter covered the use of correlations to measure the relationship between random variables. Joint distributions of discrete random variables were introduced first. The covariance of two random variables was defined as the expected value of the product of their deviations from their respective means, and their correlation was defined as their covariance divided by the product of their standard deviations. Among these two measures of the relationship between random variables, we emphasized the correlation because it is a unit-free number between −1 and 1, and it is somewhat easier to interpret. We saw correlation arrays for two or more random variables, which display the correlations among all pairs of these random variables, and always have 1's along the diagonal from top-left to bottom-right.

We studied some general formulas for linear functions of random variables. When the means, standard deviations, and pairwise correlations for a set of random variables are known, we learned general formulas for computing the expected values, variances, standard deviations, and covariances of any linear functions of these random variables. We also saw the simpler special form of these formulas when the underlying random variables are independent.

We learned how correlations are estimated from sample data by the CORREL function. The Excel function CORREL was introduced in this chapter. For a more precise assessment of the accuracy of the sample average as an estimate for an unknown expected value, we introduced Normal distributions and the central limit theorem. We learned that a sample average, as a random variable, has a standard deviation that is inversely proportional to the square root of the sample size. We then saw how to compute a 95% confidence interval for the expected value of a random variable, using simulation data.

REFERENCES

Ale B. (1991), "Acceptability Criteria for Risk and Plant Siting in the Netherlands," VIII Meeting 3ASI: Safety Reports in EC, Milan, France, September 18th-19th.

Ang, A. H-S., Tang, W. H. (1984), Probability Concepts in Engineering Planning and Design, Volume II − Decision, Risk, and Reliability," John Wiley & Sons, New York.

Anon (1980), Risk Analysis, An Official Publication of the Society for Risk Analysis, Plenum Press, New York NY.

Apostolakis, G., B. J. Garrick and D. Okrent (1983), "On Quality, Peer Review, and the Achievement of Consensus in Probabilistic Risk Analysis," Nuclear Safety, Vol. 24, No. 6, November/December, pp. 792-800.

Chang, S. H., J. Y. Park and M. K. Kim (1985), "The Monte-Carlo Method Without Sorting for Uncertainty Propagation Analysis in PRA," Reliability Engineering, Vol. 10, pp. 233-243.

Committee on Public Engineering Policy (1972), Perspectives on Benefit-Risk Decision Making, National Academy of Engineering, Washington DC.

Gorden, J. E. (1978), Structures: Or Why Things Don't Fall Down, Da Capo Press, New York, NY.

Kaplan, S. and B. J. Garrick (1981), "On the Quantitative Definition of Risk," Risk Analysis, Vol. 1, No. 1, pp. 11- 27.

Kyburg, H. E. Jr. and H. E. Smokler (1964), Studies in Subjective Probability, John Wiley & Sons, New York, NY.

Lichtenberg, J. and D. MacLean (1992), "Is Good News No News?", The Geneva Papers on Risk and Insurance, Vol. 17, No. 64, July, pp. 362-365.

March, J. G. and Z. Shapira (1987), "Managerial Perspectives on Risk and Risk Taking," Management Science, Vol. 33, No. 11, November, pp. 1404-1418.

Morgan, M. G. and M. Henrion (1990), Uncertainty: Guide to dealing with Uncertainty in Quantitative Risk and Policy Analysis, Cambridge University Press, Cambridge, Great Britain.

Roush, M. L., Modarres, M., Hunt, R. N., Kreps, and Pearce, R. (1987), Integrated Approach Methodology: A Handbook for Power Plant Assessment, SAND87-7138, Sandia National Laboratory, Albuquerque, New Mexico.

Rowe, W. D. (1994). "Understanding Uncertainty," Risk Analysis, Vol. 14, No. 5, pp. 743-750.

Schlager, N., ed. (1994), When Technology Fails: Significant Technological Disasters, Accidents, and Failures of the Twentieth Century, Gale Research, Detroit, Michigan.

Shafer, G. and J. Pearl, ed. (1990), Readings in Uncertain Reasoning, Morgan Kaufmann Publishers Inc., San Mateo CA.

Slovic, P. (1987), "Perception of Risk," Science, Vol. 236, pp. 281-285.

Slovic, P. (1993), "Perceived Risk, Trust, and Democracy," Risk Analysis, Vol. 13, No. 6, pp. 675- 682.

Wang, J. X. (1991), "Fault Tree Diagnosis Based on Shannon Entropy," Reliability Engineering and System Safety, vol. 34, pp. 143-167.

Wahlstrom, B. (1994), "Models, Modeling and Modellers: an Application to Risk Analysis," European Journal of Operations Research, Vol. 75, No. 3, pp.477-487.

Wenk, E. et. al. (1971), Perspectives on Benefit-Risk Decision Making, The National Academy of Engineering, Washington DC.

Zadeh, L. A. and J. Kacprzyk, ed. (1992), Fuzzy Logic for the Management of Uncertainty, John Wiley and Sons, New York, NY.

5

Performing Engineering Predictions

When two engineering variables are related, it is possible to predict an engineering variable from another variable with better than chance accuracy. This chapter describes how these predictions are made and what can be learned about the relationship between the variables by developing a prediction equation. Given that the relationship is linear, the prediction problem becomes one of finding the straight line that best fits the data. Since the terms "regression" and "prediction" are synonymous, this line is called the regression line. The mathematical form of the regression line predicting Y from X is: $Y = bX + A$.

5.1 CASE STUDY: A CONSTRUCTION PROJECT

We have emphasized the importance of building an appropriate relationship among random variables in a simulation model. The concept of correlation was developed by statisticians as a way to measure relationships among random variables. But correlation does not give a complete measure of the relationship between any pair of random variables in general, and correlation is not the only way that we should build relationships among random variables in our models.

A more general way to build relationships among random variables in a spreadsheet model is by making the formula for one random variable depend on the realized value of other random variables. In this chapter, we focus such models, with particular emphasis on models in which the expected value of one random variable may depend on the realized value of another random variable.

The word "depend" here deserves some careful discussion, because it has two possible meanings. Two random variables are statistically dependent whenever they are not independent, meaning only that learning about one random variable could cause you to revise your beliefs about the other random variable. In this statistical sense, the dependence of two random variables does not imply that they have any cause-and-effect relationship in which one influences the other.

Statistical dependence could also arise, for example, when two random variables are both influenced by some other random factor. But in spreadsheet models, we build cause-and-effect relationships among cells in their formulas, which gives us another concept of formulaic dependence. A cell is formulaically dependent on all the cells to which it refers in its formula, and on all the cells to which these cells refer, and so on.

Let's look at an example. A dam construction project has two stages. Let X denote the cost of the first s tage and let Y denote the cost of the second stage (both in millions of dollars). Suppose that the joint distribution of these costs is as follows:

	$Y = 30$	$Y = 32$	$Y = 34$
$X = 70$	0.1	0.1	0
$X = 75$	0.2	0.3	0
$X = 80$	0	0.2	0.1

We need to compute the expected value of Y, the conditionally expected value of **Y** given each of the possible values of X. Also, if we learned that **X** > 72 (that is, **X** may equal 75 or 80, but not 70), then what would be our conditionally expected value of **Y** given this new information?

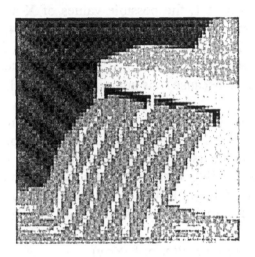

Figure 5.1 A dam construction project.

The correlation that we studied in the preceding chapter gives us an easy way to build statistical dependence among random variables without either random variable formulaically depending on the other. But we should also use formulaic dependence as a more general and versatile method for building appropriate statistical dependence among random variables.

We have learned to make random variables in spreadsheets by formulas in which an inverse cumulative function (like NORMINV) has a RAND as its first parameter. The inverse cumulative function then has other distributional parameters (commonly the mean and standard deviation) that define the specific shape of the probability distribution. The simplest way to build formulaic dependence among random variables to set these distributional parameters in one random variable as formulas that refer to the realized value of another random variable.

5.2 THE LAW OF EXPECTED POSTERIORS

For the case study in Section 5.1, the possible values of **X** are 70, 75, and 80, while the possible values of **Y** are 30, 32, and 34. The table of joint probabilities $P(X=x \ \& \ Y=y)$ discussed in Section 5.1 is now shown in the range A3:D6 of Figure 5.2. We use "&" here to denote intersection of events, because $\cap$ is unavailable in Excel.

For any number x that is a possible value of X, the marginal probability $P(X=x)$ is computed by summing the joint probabilities $P(X=x \ \& \ Y=y)$ over all numbers y that are possible values of Y. That is, the marginal probability distribution for **X** is found by summing across each row of the joint-probability table, as shown in cells F4:F6 of Figure 5.2. The mathematical formula is

$$P(X=x) = \Sigma_y P(X=x \ \& \ Y=y)$$

Similarly, the marginal probabilities $P(Y=y)$ are computed by summing down each column of the joint-probability table, as shown in cells B8:D8 of Figure 5.2.

For all numbers x and y that are possible values of **X** and Y, the conditional probability $P(Y=y|X=x)$ is computed by the basic formula

$$P(Y=y|X=x) = P(X=x \ \& \ Y=y)/P(X=x)$$

This formula is applied in the range I4:K6 in Figure 5.2. Comparing each row of conditional probabilities in I4:K6 with the corresponding row of joint probabilities in B4:D6, you can see that the conditional probabilities in the row are respectively proportional to the joint probabilities in the same row, where the constant of proportionality is what it takes to make the conditional probabilities sum to

1 across the row. For example, the top row of conditional probabilities given X=70 in I4:K4

$$P(Y=30|X=70) = 0.1, P(Y=250|X=90) = 0.1, P(Y=300|X=90) = 0$$

which are proportional to the corresponding joint probabilities (0.1, 0.1, 0) in B4:D4.

	A	B	C	D	E	F
1	JointProbys P(X=x&Y=y)					
2		y=				
3	x= \	30	32	34		P(X=x)
4	70	0.1	0.1	0		0.2
5	75	0.2	0.3	0		0.5
6	80	0	0.2	0.1		0.3
7		P(Y=y)				sum
8		0.3	0.6	0.1		1
9						
10	X	Y				
11	75.5	31.6	Expected Value			
12	3.5	1.2	Standard deviation			
13						
14	FORMULAS FROM RANGE A1:F12					
15	F4. =SUM(B4:D4)					
16	F4 copied to F4:F6.					
17	B8. =SUM(B4:B6)					
18	B8 copied to B8:D8.					
19	F8. =SUM(B4:D6)					
20	A11. =SUMPRODUCT(A4:A6,F4:F6)					
21	B11. =SUMPRODUCT(B3:D3,B8:D8)					
22	A12. = SUMPRODUCT((A4:A6-A11)^2,F4:F6)^0.5					
23	B12. = SUMPRODUCT((B3:D3-B11)^2,B8:D8)^0.5					

Figure 5.2 The law of expected posteriors.

If we learned that the value of **X** was some particular number x, then our conditionally expected value of **Y** given that X=x would be

$$E(Y|X=x) = \Sigma_y P(Y=y|X=x)*y$$

where the summation is over all number y that are possible values of Y. In our example, the conditionally expected value of **Y** given X=70 is

$$E(Y|X=70) = 0.5*30 + 0.5*32 + 0 * 34 = 31.00$$

Similarly, the conditionally expected value of **Y** given X=75 in this example is

$$E(Y * X=75) = 0.4*30 + 0.6*32 + 0 * 34 = 31.20$$

The conditionally expected value of **Y** given X=80 is

$$E(Y * X=80) = 0*30 + 0.667*32 + 0.333*34 = 32.67$$

These conditionally expected values are computed by SUM-PRODUCT formulas in cells M4:M6 in Figure 5.2. If we are about to learn **X** but not Y, then our posterior expected value of **Y** after learning **X** will be either 31.00 or 31.20 or 32.67, depending on whether **X** turns out to be 70, 75, or 80. These three possible value of **X** have prior probability 0.2, 0.5, and 0.3 respectively. So our prior expected value of the posterior expected value of **Y** that we will apply after learning **X** is

$$E(E(Y*X)) = 0.2*31.00 + 0.5*31.20 + 0.3*32.67 = 31.6$$

as shown in cell M9 of Figure 5.2. But before learning X, our expected value of **Y** is the same number, as shown in cell C11 of Figure 5.2:

$$E(Y) = 0.3 *30 + 0.6*32 + 0.1*34 = 31.6$$

More generally, the law of expected posteriors says that, before we observe any random variable X, our expected value of another random variable **Y** must be equal to the expected value of what we will think is the expected value of **Y** after we observe X. The formula for the expected posterior law is be written:

$$E(Y) = \Sigma_x P(X=x)*E(Y^*X=x)$$

(Here the summation is over all x that are possible values of X.) Or, as we may write this formula more briefly as $E(Y) = E(E(Y^*X))$ which may be briefly summarized as "the expected value of the expected value is equal to the expected value." You can find a proof of this law of expected posteriors in a book on probability and statistics.

As another application of this expected-posterior law, let A denote any event, and define **W** to be a random variable such that **W** = 1 if the event A is true, but **W** = 0 if the event A is false. Then

$$E(W) = P(A) * 1 + (1-P(A)) * 0 = P(A).$$

That is, the expected value of this random variable **W** is equal to the probability of the event A.

Now suppose that **X** is any other random variable such that observing the value of **X** might cause us to revise our beliefs about the probability of A. Then the conditionally expected value of **W** given the value of **X** would be similarly equal to the conditional probability of A given the value of X. That is, for any number x that is a possible value of the random variable X,

$$E(W|X=x) = P(A|X=x).$$

Thus, the general equation

$$E(W) = \Sigma_x P(X=x)*E(W|X=x)$$

gives us the following probability equation:

$$P(A) = \Sigma_X P(X=x)*P(A|X=x)$$

which may be written more briefly as $P(A) = E(P(A*X))$. This equation says that the probability of A, as we assess it given our current information, must equal the current expected value of what we would think is probability of A after learning the value of X. Learning **X** might cause us to revise our assessment of the probability of A upwards or downwards, but the weighted average of these possible revisions, weighted by their likelihoods, must be equal to our currently assessed probability P(A).

In our example, let A denote the event that Y=32. So before **X** is learned, the prior probability of A is

$$P(A) = P(Y=32) = 0.6,$$

as shown in cell C8 of Figure 5.2. Then the posterior probability of A given the value of **X** would be either 0.5 or 0.6 or 0.667, depending on whether **X** is 70 or 75 or 80, as shown in cells J4:J6. But the respective probabilities of these three values of **X** are 0.2, 0.5, and 0.3, as shown in cells 4:F6. So before we learn X, our prior expected value of the posterior probability of A given **X** is

$$E(P(A|X)) = 0.2*0.5 + 0.5*0.6 + 0.3*0.667 = 0.6$$

as shown in cell J11.

5.3 LINEAR REGRESSION MODELS

Engineers frequently collect paired data in order to understand the characteristics of an object or the behavior of a system. The data may indicate a spatial profile (snowfall in different cities and states) or a time history (vibration acceleration versus time). Or the data may indicate cause-and-effect relationships (for example, force exerted by a spring versus its displacement from its equilibrium position) or system output as a function of input (yield of manufacturing process as a function of its process capability). Such relationships are often developed graphically, by plotting the data in a particular way. Mathematical expressions that capture the relationships shown in the data can then be developed.

We now consider a special form of formulaic dependence among random variables that is probably assumed by statisticians more often than any other: the linear regression model. In a linear regression model, a random variable Y is made dependent on other random variables $(X_1, ..., X_k)$ by assuming that Y is a linear function of these other random variables $(X_1, ..., X_k)$ plus a Normal error term. A Normal error term is a Normal random variable that has mean 0 and is independent of the values of the other random variables $(X_1, ..., X_k)$. To define such a linear 1 K regression model where Y depends on K other random variables, we must specify K+2 constant parameters: the coefficients for each of the K explanatory random variables in the linear function, the Y-axis intercept of the linear function, and the standard deviation of the Normal error term.

An example of a linear regression model is shown in Figure 5.3, where a random variable Y, the mechanical wear, in cell B12 is made to depend on another random variable X, the spring force applied, in cell A12. The value 3 in cell B4 is the coefficient of X in the linear function, the value -10 in cell B6 is the Y-axis intercept of the linear function, and the value 8 in cell B8 is the standard deviation of the Normal error term. So with these parameters in cells B4, B6, and B8, the formula for Y in cell B12 is

=\$B\$6+\$B\$4*A12+NORMINV(RAND(),0,\$B\$8)

The value of the X in cell A12 here is generated by a similar formula

=\$A\$4+NORMINV(RAND(),0,\$A\$6)

where cell A4 contains the value 10 and cell A6 contains the value 2. That is, the random variable **X** here has no random precedents, and it is simply the constant 10 plus a Normal error term with standard deviation 2.

	A	B	C	D	E	F	G	H	I	J	K	L
1	Parameters			Computed from parameters								
2	for X	for Y given X		E(Y)	Covar (X, Y)							
3	E(X)	X coefficient		20	12							
4	10	3		Stdev (Y)	Correl(X, Y)							
5	Stdev(X)	Intercept		10	0.6							
6	2.1											
7		Std error										
8		8										
9												
10	Data			Computed from data								
11	X	Y		Estimates for X						Estimates for Y		
12	11.69	15.45		E(X)						E(Y)		
13	10.39	17.03		9.799						17.167		
14	9.91	25.2		Stdev(X)						Stdev(Y)		
15	8.44	10.66		2.418						6.306		
16	8.39	23.99								Correl(X, Y)		
17	8.1	14.83								0.4038		
18	9.44	15.37										
19	9.85	19.37										
20	14.29	34.83		FORMULAS								
21	11.5	10.17		E13. =AVERAGE(12.A31)						J13. = AVERAGE(B12:B31)		
22	8.45	6.21		E15. =STDEV(A12:A31)						J15. = STDEV(B12:B31)		
23	15.84	17.43								J17. = CORREL(A12:A31, B12:B31)		
24	8.6	16.16		E3. =B6+B4*A4								
25	6.46	14.99		F3. =B4*A6^2								
26	11.87	16.89		E5. =((A6*B4)^2+B8^2)^0.5								
27	6.8	12.21		F5. =F3/(A6+E5)								
28	10.56	15.29										
29	8.86	23.13										
30	10.16	21.25										
31	6.41	12.86										

Figure 5.3 A linear regression model for mechanical wear versus spring force applied.

But adding a constant to a Normal random variable yields a Normal random variable with mean increased by the amount of this constant. So **X** in cell A12 is a Normal random variable with mean 10 and standard deviation 2, and the formula in cell A12 is completely equivalent to

=NORMINV(RAND(),$A$4,$A$6)

Similarly, the formula for **Y** in cell B12 could have been written equivalently as

=NORMINV(RAND(),$B$6+$B$4*A12,$B$8)

which is shorter but harder to read.

The formulas from A12:B12 in Figure 5.2 have been copied to fill the range A12:B31 with 20 independent pairs that have the same joint distribution. That is, in each row from 13 to 31, the cells in the A and B column are drawn from the same joint distribution as **X** and **Y** in A12:B12. In each of these rows, the value of the B-cell depends on the value of the A-cell according to the linear regression relationship specified by the parameters in $B$3:$B$8, but the values in each row are independent of the values in all other rows.

The fundamental problem of statistical regression analysis is to look at statistical data like the 20 pairs in A12:B31, and guess the linear regression relationship that underlies the data. Excel offers several different ways of doing statistical regression analysis. In particular, with the Excel's Analysis ToolPak added in, you should try using the menu command sequence

Tools>DataAnalysis>Regression.

If you try recalculating the spreadsheet in Figure 5.3, so that the simulated **X** and **Y** data is regenerated in cells A12:B31, you will see the regression estimates in F12:F18 varying around the true underlying parameters that we can see in B3:B8. More sophisticated regression-analysis software (such as that in Excel's Analysis Tool-Pak) will generate 95% confidence intervals for the regression parameters around these estimates.

Regression analysis in F12:F18 here only assesses the parameters of the conditional probability distribution of **Y** given X. So to complete our estimation of the joint distribution of **X** and Y, we need to estimate the parameters of the marginal probability distribution of X. If we know that **X** is drawn from a Normal probability distribution but we do not know its mean and standard deviation, then we can use the sample average and the sample standard deviation of our **X** data (as shown in cells E13 and E15 of Figure 5.3) to estimate the mean and standard deviation parameters of the Normal distribution that generated our **X** data.

Statistical regression analysis in general does not require any specific assumptions about the probability distributions that generated the explanatory **X** data. But if the explanatory **X** is a Normal random variable (or, when K > 1, if the explanatory **X** variables are Multivariate Normal), then the joint distribution of **X** and **Y** together is Multivariate Normal. Thus our **X** and **Y** random variables in A12:B12, for example, are Multivariate Normal random variables.

The difference between the treatment of Multivariate Normals here and in Chapter 4 is the parametric form that we use to characterize these Multivariate Normal random variables. Here the parameters are the mean E(X) and standard deviation Stdev(X), which characterize the marginal distribution of X, and the Intercept, Coefficient, and StdError parameters of the regression relationship

Y = Intercept + Coefficient * **X** +NORMINV(RAND(),0,StdError)

which characterize the conditional distribution of **Y** given X. But in Chapter 4 we parameterized the Multivariate Normal random variables by their individual means, standard deviations, and pairwise correlations. In cases like this, where we have only one explanatory variable **X** in the regression model, the other parameters of the Multivariate-Normal distribution of **X** and **Y** can be computed from the regression parameters by the equations:

E(Y) = Intercept + Coefficient * E(X)

Stdev(Y) = ((Coefficient) * Stdev(X))^2 + StdError^2)^0.5

Correl(X,Y) = (Coefficient) * Stdev(X)'Stdev(Y)

These equations are applied to compute these parameters in E2:F5 of Figure 5.3. These computed parameters are then used with CORAND (in G5:H5) to make a simulation of **X** and **Y** in cells G7:H7 that is completely equivalent to the simulation that we made with regression in cells A12:B12. That is, no statistical test could distinguish between data taken from cells A12:B12 and data taken from cells G7:H7.

Notice that the standard deviation of Y, as computed in cell E5 of Figure 5.3, is larger than the standard error of the regression, which has been given in cell B8 (10 > 8). The standard error of the regression is the conditional standard deviation of Y given X. Recall that such a conditional standard deviation should be generally smaller than the unconditional standard deviation, because the standard deviation is a measure of uncertainty and learning X tends to reduce our uncertainty about the dependent variable Y.

5.4 REGRESSION ANALYSIS AND LEAST SQUARED ERRORS

Regression analysis is a powerful statistical method that helps engineers understand the relationship between engineering variables. Here we will show how the estimates of a statistical regression analysis are determined from the data. Consider the quality index data for Processes 1 and 2 shown in cells B2:C11 of Figure 5.4, and let us try to estimate a regression model in which the quality index of Process 2 depends on the quality index of Process 1 in each year.

We start by searching for a linear function that can be used to predict the quality index of Process 2 as a function of the quality index of Process 1. In Figure 5.4, the values of cells A14 and B14 will be interpreted as the intercept and coefficient of this linear function. Then, based on the 1990 quality index of Process 1 in cell B2, our linear estimate of estimate of Process 2's quality in 1990 can be computed by the formula

=$A$14+$B$14*B2

which has been entered into cell D2. Then copying cell D2 to D2:D11 gives us a column of linear estimates of Process 2's quality index based only on the quality index of Process 1 in the same year.

The numerical value of these estimates will depend, of course, on the intercept and coefficient that we have entered into cells A14 and B14. For example, if we enter the value 0 into cell A14 and the value 1 into cell B14, then our linear estimate of Fund 2's growth will be just the same as Fund 1's growth for each year. Unfortunately, no matter what intercept and coefficient we may try, our linear estimates in D2:D11 will be wrong (different from C2:C11) in most or all the years for which we have data. But we can at least ask for an intercept and coefficient that will generate linear estimates that have the smallest overall errors, in some sense.

The overall measure of our errors that statisticians use in regression analysis is an adjusted average the squared errors of our linear estimates. The squared error of the linear estimate in 1990 is computed in cell E2 of Figure 5.4 by the formula

$$=(D2-C2)^\wedge 2$$

and this formula in E2 has then been copied to E2:E11. The adjusted average squared error is computed in cell E14 by the formula

$$=SUM(E2:E11)/(COUNT(E2:E11)-COUNT(A14:B14))$$

That is, the adjusted average differs from the true average in that we divide, not by the number of data points, but by the number of data points minus the number of linear parameters that are we have to find in the linear function. This denominator (number of data points minus number of parameters in the linear function) is called the degrees of freedom in the regression.

Now let us ask Solver to minimize this adjusted average squared error in cell E14 by changing the parameters of the linear function in cells A14:B14. The resulting optimal solution that Solver returns for this optimization problem is shown in cells A14:B14 of Figure 5.4 (an intercept of -0.1311 and a coefficient of 1.1305). That this optimal solution is the same as the intercept and coefficient that the regression analysis returns, because statistical regression analysis is designed to find the linear function that minimizes the adjusted average squared error in this sense.

Cell E16 in Figure 5.4 computes that square root of the adjusted average squared error, with the formula

$$=E14^0.5$$

Notice that the result of this formula in cell E16 is the same as the standard error of the regression is that regression analysis. That is, the estimated standard error of the regression (or the estimated standard deviation of the Normal error term in the regression model) is the square root of the minimal adjusted-average squared error.

If we had used the simple average in cell E14 (dividing by the number of data points, instead of dividing by the number of degrees of freedom) then Solver would have still given us the same optimal intercept and coefficient in A14:B14 when we asked it to minimize E14. But the square root of the average squared error would have been slightly smaller. Regression analysis uses this adjusted average squared error (dividing by the number of degrees of freedom) because otherwise our ability to adjust these linear parameters to fit our data would bias downwards our estimate of the variance of the Normal errors in the regression.

In cells E19:F19 of Figure 5.4, this statistically estimated regression model is applied to make a simulation model for forecasting the quality index of Processes 1 and 2 in future years. The quality index of Process 1 is simulated in cell E19 by the formula

$$=NORMINV(RAND(),AVERAGE(B2:B11),STDEV(B2:B11))$$

	A	B	C	D	E	F	G
		Process 1	Process 2				
		quality	quality	Linear estimate			
1		index	index	of 2	Squared error		
2	1990	1.22	1.28	1.2427	0.00141		
3	1991	1.09	1.20	1.0962	0.01036		
4	1992	0.92	0.83	0.9033	0.00509		
5	1993	1.37	1.48	1.4226	0.00321		
6	1994	1.01	0.96	1.0053	0.00221		
7	1995	1.15	1.22	1.1645	0.00263		
8	1996	1.26	1.15	1.2921	0.01982		
9	1997	1.25	1.17	1.2856	0.01309		
10	1998	0.93	0.96	0.9163	0.00153		
11	1999	1.19	1.31	1.2198	0.00759		
12							
13	Intrcept	Coefficient			Adjusted avg sq err		
14	-0.1311	1.1305			0.00837		
15					Std err		
16	SOLVER: minimize E14 by changing A14:B14				0.0915		
17							FORMULAS
18			PROCESS2|PROCESS1			D2. =$A$14+$B$14*B2	
19			Regressn Array ... 7 Rows x 1 Columns			E2. =(D2-C2)^2	
20			X(1) Coefficient			E2:D2 copied to E2:D11	
21			1.1305		E14. =SUM(E2:E11)/(COUNT(E2:E11)-COUNT(A14:B14))		
22		PROCESS1 Mean	Intercept			E16. =E14^0.5	
23		1.1375	-0.1311			B23. =AVERAGE(B2:B11)	
24		PROCESS1 Stdev	Std Error			B25. =STDEV(B2:B11)	
25		0.1520	0.0915		C19:C25. {=REGRESSN(B2:B11,C2:C11)}		
26					B29. =B23+NORMINV(RAND(),0,B25)		
27	Simulation model:				C29. =C23+C21*B29+NORMINV(RAND(),0,C25)		
28		Process 1	Process 2				
29	SimTable	0.968	1.010				
30	0	1.1516854	1.09708789				
31	0.005	1.3411713	1.53527553				
32	0.01	0.9775424	1.00311015				
33	0.015	1.1130184	1.16980256				
34	0.02	0.9721749	0.98816738				
35	0.025	1.1424716	1.0833063				
36	0.03	1.184588	1.28823462				
37	0.035	0.8334201	0.72680957				
38	0.04	1.0769892	1.07926567				
39	0.045	0.9562577	0.92133086				
40	0.05	1.2364224	1.15549292				
41	0.055	1.5691487	1.48982434				
42	0.06	0.9799988	0.93468675				
43	0.065	1.1733234	1.06256496				
44	0.07	1.1430508	1.05346126				
45	0.075	1.132406	1.21435169				

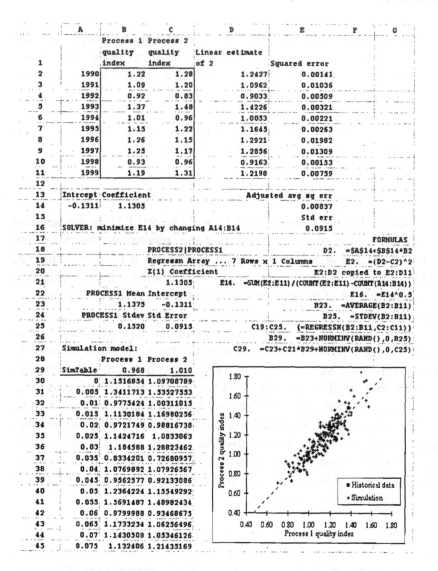

Figure 5.4 Regression analysis of process quality index.

Then the quality index of Process 2 is simulated in cell F19 by the formula

$$=A21+A19*E19+NORMINV(RAND(),0,A23)$$

using the intercept, coefficient, and standard error from the analysis. A simulation table containing 200 independent recalculations of cells E19:F19 has been made in this spreadsheet and the results are shown in the scatter chart in the lower right of Figure 5.4. The optimal linear function for estimating Process 2 from Process 1 is also shown by a dashed line in that chart.

Excel provides a regression feature within the Analysis Toolpak. This feature will fit a straight line (called a *regression* line) to a set of data using the method of least square. The resulting output includes the coefficients of the regression line equation, the sum of the squares of the errors, and the Υ^2 value. In addition, optional output can be requested, which includes a listing of the individual error terms (called *residuals*) and a plot of the data.

Fit a straight line through the data given in Example 5.4 using the Regression feature found in the Excel Analysis Toolpak. Include a list of the residuals in the output and generate a plot of the data. Figure 5.5 shows the resulting output, once OK has been selected in the Regression dialog box. Within the computed output, we see that the Υ^2 value (0.79878) is shown in cell B5, under the heading *Regression Statistics*. Similarly, the sum of the squares of the errors (0.066932) is shown in cell C13, under the heading *SS*. This value is labeled *Residual*.

The coefficients in the regression equation are given in cells B17 and B18, under the heading *Coefficients*. Thus, the y-intercept (-0.13113) is shown in cell B17, and the slope (1.13049) in cell B18. Also, the predicted y-values are shown in cells B25 through B29, and corresponding error terms (residuals) in cells C25 through C29. Excel can perform two variable or multiple regressions, and provides several useful analyses of residuals. It does want the variables in a particular order, with the Y or dependent variable first or to the left, and all the X or independent variables in columns to the right. It will not allow you to select non-adjacent columns for Y and X. To perform a regression, click on the TOOLS button on the TOOL-

BAR, and click on DATA ANALYSIS on the pull-down menu. That opens a DIALOG BOX and you can scroll down the list to click REGRESSION. Another BOX is opened, where you specify the Y or dependent variable. Specify as a column such as C4:C20; the data for Y are in rows 4-20 of column C, or click on the spreadsheet icon to select the variable. Then move the cursor to the X variables box and specify one or more X or independent variables. It helps if your X variables are all adjacent. For example, you can specify all the variables in columns D through G with the address D4:G20. Then click the boxes to mark which analyses you want, such as a column for the residuals or standardized residuals, and for plots of the residuals. You can suppress the intercept if you want by clicking that box. The default for the location of the regression results is a new "ply" or sheet in the spreadsheet, so it will write the results and place the graphs on a new worksheet.

	A	B	C	D	E	F	G	H	I
1	SUMMARY OUTPUT								
2									
3	*Regression Statistics*								
4	Multiple R	0.893744845							
5	R Square	0.798779848							
6	Adjusted R Sqr	0.773627329							
7	Standard Error	0.091468587							
8	Observations	10							
9									
10	ANOVA								
11		df	SS	MS	F	Significance F			
12	Regression	1	0.265698777	0.265698777	31.73745	0.000489665			
13	Residual	8	0.06693282	0.008366502					
14	Total	9	0.332630796						
15									
16		Coefficients	Standard Error	t Stat	P-value	Lower 95%	Upper 95%	Lower 95.0%	Upper 95.0%
17	Intercept	-0.13112963	0.230021233	-0.570076199	0.584276	-0.661559887	0.39930063	-0.661559887	0.399300627
18	X Variable 1	1.130494615	0.200606819	5.635374812	0.00049	0.667894162	1.59309507	0.667894162	1.593095069
19									
20									
21									
22	RESIDUAL OUTPUT								
23									
24	Observation	Predicted Y	Residuals						
25	1	1.242711108	0.037601524						
26	2	1.09618244	0.101804691						
27	3	0.983304698	-0.071325632						
28	4	1.422613609	0.056611685						
29	5	1.005256814	-0.046998085						
30									

Figure 5.5 Use the regression feature in Excel.

5.5 THE THEORY OF CONSTRAINTS

The *theory of constraints* (or TOC as it is called) is a relatively recent development in the practical aspect of making organizational decisions in situations in which constraints exist. The theory was first described by Dr. Eliyahu M. Goldratt in his novel, *The Goal.* In many organizations, TOC and TOC logic form major portions of that organization's philosophy of continuous improvement.

The *theory of constraints* has been used at three different levels:

1) *Production Management* - TOC was initially applied here to solve problems of bottlenecks, scheduling, and inventory reduction.

2) *Throughput analysis* - Application of TOC has caused a shift from cost-based decision making to decision making based on continuous improvement of processes in which system throughput, system constraints, and statistically determined protective capacities at critical points are key elements.

3) *Logical Processes* - This third level is the general application of TOC reasoning to attack a variety of process problems within organizations. TOC logic is applied to identify what factors are limiting an organization from achieving its goals, developing a solution to the problem, and getting the individuals in the process to invent the requisite changes for themselves.

A constraint is anything in an organization that limits it from moving toward or achieving its goal. Of course, this assumes that an appropriate goal has been defined. For most business organizations the goal is to make money now as well as in the future. There are two basic types of constraints: physical constraints and non-physical constraints. A physical constraint is something like the physical capacity of a machine. A non-physical constraint might be something like demand for a product, a corporate procedure, or an individual's paradigm for looking at the world.

The steps in applying TOC are as follows:

1. Identify the system's constraints. Of necessity this included prioritization so that just the ones that really limit system progress toward the goal.

2. Decide how to exploit the system's constraints. Once we have decided how to manage the constraints within the system, how about the majority of the resources that are not constraints? The answer is that we manage them so that they just provide what is needed to match the output of the constrained resources. We NEVER let them supply more output than is needed because doing so moves us no closer to the goal.

3. Subordinate everything else to the above decision in Step 2. Since the constraints are keeping us from moving toward our goal, we apply all of the resources that we can to assist in breaking them. Constraints are not acts of God. In practically all cases their limiting impact can be reduced or eliminated.

4. Elevate the system's constraints. If we continue to work toward breaking a constraint (also called elevating a constraint) at some point the constraint will no longer be a constraint. The constraint will be broken.

5. If the constraint is broken, return to Step 1. When that happens, there will be another constraint, somewhere else in the system that is limiting progress to the goal.

The process must be reapplied, perhaps many times. It is very important not to let inertia become a constraint. Most constraints in organization are of their own making. They are the entrenched rules, policies, and procedures that have developed over time. Many times, when we finally break a constraint, we do not go back and review and change the rules and policies that caused the constraint initially. Most constraints in organizations today are policy constraints rather than physical constraints.

For a manufacturing organization, with the goal being to make money now as well as in the future, TOC defines three operational measurements that measure whether operations are working toward that goal. They are:

Throughput: The rate at which the system generates money through sales. This is considered to be the same as Contribution Margin (selling price -- cost of raw materials). Labor costs are considered to be part of Operating Expense rather than throughput.

Inventory: All the money the system invests in things it intends to (or could) sell. This is the total system investment, which includes not only conventional inventory, but also buildings, land, vehicles, plant, and equipment. It does not include the value of labor added to Work-In-Process inventory.

Operating Expense: All the money the system spends in turning Inventory into Throughput. This includes all of the money constantly poured into a system to keep it operating, such as heat, light, scrap materials, depreciation, etc.

The following four measurements are used to identify results for the overall organization:

Net Profit = Throughput - Operating Expense

Return on Investment (ROI)
    = (Throughput - Operating Expense) / Inventory

Productivity = Throughput / Operating Expense

Turnover = Throughput / Inventory

Given the measurements as described, employees can make local decisions by examining the effect of those decisions on the organization's overall throughput, inventory, and operating expense. A decision that results in increasing overall throughput, decreasing the overall inventory, or decreasing the overall operating expense for the firm will generally be a good decision for the business.

The *theory of constraints* does away with much of cost accounting. It is clear that application of cost accounting principles (primarily the allocation of costs in order to make decisions at the local level) leads to poor management decisions at the department as well as in upper levels of the organization. In fact, TOC virtually eliminates the use of economic order quantities (EOQ), production lot sizes, deriving product costs, setting prices, determining productivity measures, and the use of performance incentives.

Most individuals will readily see the use for *the theory of constraints* in the improvement of production scheduling or in improving manufacturing. This is simply incorrect. Although it is true that the *theory of constraints* provides us with simple examples in the manufacturing environment, TOC is truly applicable to any process in any organization. This includes universities, hospitals, service providers of all varieties, government and, of course, manufacturing.

5.6 SUMMARY

Regression analysis is a method for exploring the relationship between a response variable and one or more explanatory variables. Regression techniques are used in areas of business, engineering, medicine, scientific, and social research. If you're a software consultant or developer, most likely your firm develops applications for collection of data. The firms with rates more than twice the average, are those that can also make useful inferences about the data collected by their systems.

The basic multiple regression model assumes a linear relationship between x_i and y, with superimposed noise (error) e

$$Y = b_0 + b_1x_1 + b_2x_2 + b_3x_3 + e \qquad (5.1)$$

Using certain statistical assumptions about Eq. (5.1), we can "fit" a regression equation

$$\underline{y} = \underline{b}_0 + \underline{b}_1 x_1 + \underline{b}_2 x_2 + \underline{b}_3 x_3 \tag{5.2}$$

as an estimator of the true relationship between the x_is and y

$$y = \beta_0 + \beta_1 x_1 + \beta_2 x_2 + \beta_3 x_3 \tag{5.3}$$

The roots of the general linear model surely go back to the origins of mathematical thought, but it is the emergence of the theory of algebraic invariants in the 1800s that made the general linear model, as we know it today, possible. The theory of algebraic invariants developed from the groundbreaking work of 19th century mathematicians such as Gauss, Boole, Cayley, and Sylvester. The theory seeks to identify those quantities in systems of equations which remain unchanged under linear transformations of the variables in the system. Stated more imaginatively (but in a way in which the originators of the theory would *not* consider an overstatement), the theory of algebraic invariants searches for the eternal and unchanging amongst the chaos of the transitory and the illusory. That is no small goal for any theory, mathematical or otherwise.

The development of the linear regression model in the late 19th century, and the development of correlational methods shortly thereafter, are clearly direct outgrowths of the theory of algebraic invariants. Regression and correlational methods, in turn, serve as the basis for the general linear model. Indeed, the general linear model can be seen as an extension of linear multiple regression for a single dependent variable. Understanding the multiple regression model is fundamental to understanding the general linear model.

REFERENCES

Brombacher, A. C. (1992), <u>Reliability by design</u>, Wiley & Sons, New York, N.Y.

Center for Chemical Process Safety (1989), Process equipment reliability data, American Institute of Chemical Engineering, New York, N.Y.

Gillett, J. (1996), Hazard Study & Risk Assessment: A Complete Guide, Interpharm Press, Incorporated.

Goldratt, E. M. (1988), "Computerized Shop Floor Scheduling," International Journal of Productivity Research (Vol. 26, No. 3), March.

Goldratt, E. M. and Cox, J. (1992), The Goal, Second Revised Edition, North River Press, Croton-on-Hudson, N.Y.

Goldratt, E. M. (1987), "Chapter 1: Hierarchical Management--The Inherent Conflict," The Theory of Constraints Journal, (Vol. 1, No. 1), Avraham Y. Goldratt Institute, New Haven, CT, Oct.-Nov.

Goldratt, E. M. (1988), "Chapter 2: Laying the Foundation," The Theory of Constraints Journal, (Vol. 1, No. 2), Avraham Y. Goldratt Institute, New Haven, CT, April - May.

Goldratt, E. M. (1988), "Chapter 3: The Fundamental Measurements," The Theory of Constraints Journal, (Vol. 1, No. 3), Avraham Y. Goldratt Institute, New Haven, CT, Aug.-Sept.

Goldratt, E. M. (1989), "Chapter 4: The Importance of a System's Constraints," The Theory of Constraints Journal, (Vol. 1, No. 4), Avraham Y. Goldratt Institute, New Haven, CT, Feb.-Mar..

Goldratt, E. M. and Fox, R. E. (1986), The Race, North River Press, Croton-on-Hudson, N.Y.

Goldratt, E. M. (1990), The Haystack Syndrome, North River Press, Croton-on-Hudson, N.Y.

Goldratt, Eliyahu M. (1990), The Theory of Constraints, North River Press, Croton-on-Hudson, N.Y.

Kaplan S. and Garrick B.J. (1981), "On the quantitative definition of risk," Risk Analysis, Vol. 1(1), pp. 11-27.

Kumamoto, H. and Henley, E. J. (1995), Probabilistic Risk Assessment & Management for Engineers & Scientists, IEEE Press, Piscataway, New Jersey.

Lemons, J. (1995), Risk Assessment (Readings for the Environment Professional), Blackwell Scientific, Malden, Massachusetts.

Marsili G., Vollono C., Zapponi G.A. (1992), "Risk communication in preventing and mitigating consequences of major chemical industrial hazards," Proceedings of the 7th International Symposium on Loss Prevention and Safety Promotion in the Process Industries, Taormina, May 4th-8th, Vol. IV, 168-1 - 168-13.

Melchers, R. E. and Stewart, M. G. (1995), "Integrated Risk Assessment: Current Practice & New Directions," Proceedings of the Conference, New Castle, NSW, Australia, June 1-2.

Melchers, R. E. and Stewart, M. G. (1994), Probabilistic Risk & Hazard Assessment, Balkema (A. A.) Publishers, Netherlands.

Roush, M.L., Weiss, D., Wang J.X., (1995), "Reliability Engineering and Its Relationship to Life Extension," an invited paper presented at the 49th Machinery Failure Prevention Technology Conference, Virginia Beach, VA, April 18-20.

Society for Risk Analysis - Europe (1996), "Risk In a Modern Society Lessons from Europe," The 1996 Annual Meeting for the Society for Risk Analysis – Europe, University of Surrey, Guildford, U.K.

Vose, David (1996), Risk Analysis: A Quantitative Guide to Monte Carlo Simulation Modelling, Wiley Liss Inc., New York, N.Y.

6

Engineering Decision Variables – Analysis and Optimization

> "Two roads diverged in a yellow wood,
> And sorry I could not travel both
> And be one traveler, long I stood
> And looked down one as far as I could
> To where it bent in the undergrowth ..."

Robert Frost (1874-1963)

Analysis and simulation of decision variables will help engineers optimize their choice.

6.1 CASE STUDY: PRODUCTION PLANNING OF SNOWBOARDS

The Peak Technology Corporation will sell snowboards next winter. Peak Technology's cost of manufacturing is $20 per snowboard. Peak Technology will get $48 in revenue for every snowboard that it sells next winter. If demand is less than supply, then the snowboards that Peak Technology still has unsold in its inventory at the end of the winter will have a value of $8 each to Peak Technology. If demand is greater than Peak Technology's supply, then excess demand will be lost to other competitors.

Peak Technology's factory and shipping schedules require that all snowboards that it sells next winter must be manufactured by September of this year, some three months before the winter season begins. As this year's manufacturing begins, Peak Technology does not have any old snowboards in inventory.

The predictions about demand for snowboards next winter depend on the general weather patterns, which may be normal or cold (Figure 6.1). If next winter's weather is normal, then demand for Peak Technology's snowboards will have probability 0.25 of being below 60,000, probability 0.5 of being below 75,000, and probability 0.75 of being below 90,000. On the other hand, if next winter's weather is cold then demand for Peak Technology's snowboards have probability 0.25 of being below 80,000, probability 0.5 of being below 100,000, and probability 0.75 of being below 125,000. It is currently estimated that the probability of cold weather next winter is 0.33, and the probability of normal weather is 0.67.

Figure 6.1 Demand for snowboards depends on weather patterns.

The staff engineer assigned to analyze this production planning decision has generally assumed that Peak Technology's objective in such situations should be to maximize the expected value of its profits. In a production planning session, a marketing manager remarked that good forecasts of the coming winter's general weather would be available in November. But the manufacturing director replied that a delay of snowboard production until November could substantially increase Peak Technology's total productions costs, perhaps by $100,000 or more. A timely decision is required now!

A decision variable is any quantity that we have to choose in a decision problem. In the Peak Technology case, the principal decision variable is the quantity of snowboards that Peak Technology will produce for next winter. Peak Technology's objective is to maximize its profit, which depends both on this production quantity and on the unknown demand for Peak Technology's snowboards. The demand for Peak Technology's snowboards next winter is the principal unknown quantity or random variable in the this case.

People sometimes confuse decision variables with random variables, because in some intuitive sense they are both "unknown" when we start a decision analysis. But of course we can control a decision variable, and so we can stop being uncertain about it any time we want to make the decision. Random variables are used in decision analysis models only to represent quantities that we do not know and cannot control.

To analyze a decision problem like the Peak Technology case, we may begin by constructing a simulation model that describes how profit returns and other outcomes of interest may depend on the decision variables that we control and on the random variables that we do not control. In this model, we can try specifying different values for the decision variables, and we can generate simulation tables that show how the probability distribution of profit outcomes could depend on the values of the decision variables. Using the criterion of expected profit maximization or some other optimality criterion, we can then try to find values of the decision variables that yield the best possible outcome distribution, where "best" may be defined by some optimality criterion like expected value maximization.

Simulation is used in decision analysis to compare the probability distributions of payoff outcomes that would be generated by different decisions or strategies in a decision problem. In this chapter, we will study some general techniques for analyzing decision problems with simulation, using the Peak Technology snowboards case as our basic example. As shown by such simulation analysis in this chapter, three different general approaches may be distinguished.

In Sections 6.2, 6.3, and 6.4, we consider three different ways that simulation can used to compare different decisions. In Section 6.5, we go beyond the expected value criterion and we introduce optimality criteria for decision making with constant risk tolerance. In Section 6.5, we consider the strategic use of information.

6.2 METHOD 1: SIMULATING PAYOFFS FOR EACH ENGINEERING STRATEGY

We can set up our simulation model assuming one strategy for the decision maker and generate a large table of simulated payoffs for this strategy. Then we can make revised versions of the simulation model that assume other feasible strategies, one at a time, and similarly generate a separate table of simulated payoffs for each strategy that we want to consider.

Let us illustrate these simulation ideas in the context of the Peak Technology snowboards case. To begin, consider Figure 6.2, which applies the this simulation approach described above, that of generating a separate simulation table of payoffs for each possible decision option.

The basic parameters of the Peak Technology case (demand quartiles in each weather state, probability of cold weather, prices and costs) are summarized in the range A1:C12. The random weather pattern (with 0 for "normal" and 1 for "cold") is simulated in cell B15. Cells E16:E19 contain IF formulas that return the quartile boundaries for the conditional distribution of demand given the simulated weather pattern in B15.

If next winter's weather is normal, then demand for Peak Technology's snowboards will have probability 0.25 of being below 60,000 (Q_1), probability 0.5 of being below 75,000 (Q_2), and probability 0.75 of being below 90,000 (Q_3). Since $Q_3 - Q_2 = Q_2 - Q_1$, we know from Chapter 3 that the demand for Peak Technology snowboards is now a Normal random variable with mean $\mu = Q_2$ and standard deviation $\sigma = (Q_3-Q_1)/1.349$.

On the other hand, if next winter's weather is cold then demand for Peak Technology's snowboards have probability 0.25 of being below 80,000, probability 0.5 of being below 100,000, and probability 0.75 of being below 125,000. Since $Q_3/Q_2 = Q_2/Q_1$, again we know from Chapter 3 that the demand for Peak Technology snowboards is now a Lognormal random variable with log-mean m = $LN(Q_2)$ and log-standard-deviation s = $(LN(Q_3)-LN(Q_1))/1.349$.

Thus demand for Peak Technology snowboards is simulated in cell B16, with the quartiles listed in E16:E19. So the probability distribution of simulated demand in cell B16 depends on the simulated weather pattern in cell B15 just as Peak Technology's demand is supposed to depend on the actual weather pattern.

Cell C17 in Figure 6.2 is used to represent the quantity of snowboards that Peak Technology will produce, which is the principal decision variable in this model. So the current value of cell C17 in Figure 6.2, which is 85,000, should be interpreted as just one tentative proposal, to be evaluated in the model. Peak Technology's actual sales of snowboards will be either the demand quantity or the production quantity, whichever is smaller. So cell B18, with the formula =MIN(B16,C17) contains the sales quantity (in snowboards) that would result from the production quantity in C17 and the demand in B16.

	A	B	C	D	E	F	G	H	I
1	Peak Technologies PARAMETERS:				FORMULAS				
2	Demand distribution given weather				B15. =IF(RAND()<B8,1,0)				
3	Quartiles	Normal	Cold		E16. =IF($B$15=1,C4,B4)				
4	q1 (.25)	60000	80000		E16 copied to E16:E18				
5	q2 (.50)	75000	100000		B16. =IF(B15=0,NORMINV(RAND(),E17,(E18-E16)/1.349),EXP(NORMINV(RAND(),LN(E17),(LN(E18)-LN(E16))/1.349)))				
6	q2 (.75)	90000	125000		B18. =MIN(B16,C17)				
7					B19. =C17-B18				
8	P(cold winter)	0.33333			B20. =B11*B18+B12*B19-B10*C17				
9					B28. =B20				
10	$Cost/unit	20			B23. =AVERAGE(B29:B429)				
11	$SellingPrice	48			B24. =STDEV(B29:B429)				
12	RemainderValue	8			B25. =PERCENTILE(B29:B429,0.1)				
13					B26. =PERCENTILE(B29:B429,0.9)				
14	Simulation Model								
15	Weather cold?	0				Demand quartiles	weather		
16	Demand	100548.7			60000				
17	Production Quantity		85000		75000				
18	Sales	85000			90000				
19	Remainder	0							
20	Profit ($)	2380000							
21									
22	Profit statistics:								
23	E(Profit)	1958679							
24	Stdev	610978.7							
25	10%-level	1028872							
26	90%-level	2380000							
27	Sim'd Profit								
28	Simulation Table	2380000							
29	0	-1134523							
30	0.0025	-389872							
31	0.005	-282980							
32	0.0075	28746.88							
33	0.01	60299.92							
34	0.0125	170686.8							
35	0.015	254901							
36	0.0175	279771.2							
37	0.02	343192.8							
38	0.0225	385668							
39	0.025	449275.5							
40	0.0275	458628.1							
41	0.03	461370.8							
42	0.0325	502366.8							
43	0.035	526600.5							
44	0.0375	528861.9							

Snowboard Production Planning - Profit Profile

Figure 6.2 Developing profit/risk profile for one snowboards production strategy.

The remaining inventory of snowboards at the end of the season (which have a remainder value of $8 per snowboard) can be computed by the formula =C17-B18 in cell B19, and the final profit (in dollars) can be computed by the formula

$$=B11*B18+B12*B19-B10*C17$$

in cell B20. (Cell B10 contains the cost per snowboard, B11 contains the selling price, and B12 contains the value of snowboards in inventory at the end of the winter.)

Cell B28 in Figure 6.2 echoes the simulated profit in cell B20 (by the formula =B20), and 401 simulated values of this simulated profit have been tabulated below in cells B29:B429. The profit data has been sorted to make the cumulative risk profile shown in the Figure 6.2, which plots the percentile index in cells A29:A429 on the vertical axis and the simulated profit values in cells B29:B429 on the horizontal axis. (Because of this sorting, Figure 6.2 shows only the worst profit outcomes at the top of the table in cells B29:B44.) The expected profit for Peak Technology with the production quantity of 85,000 snowboards is estimated to be approximately $1,934,000 in cell B23 by the formula

$$=AVERAGE(B29:B429).$$

Figure 6.2 only shows us an analysis of one possible decision: that of producing 85,000 snowboards. But it is easy for us to enter other possible values of this decision variable into cell C17, and regenerate the simulation table. By entering the production quantities 80,000 through 85,000 successively into cell C17 and regenerating the simulation table in this spreadsheet once for each production quantity, we should get estimates of expected profit in cell B23 similar to the follows:

Production quantity	Average simulated profit
80,000	1,958,679
81,000	1,891,362
82,000	1,909,758
83,000	1,852,003
84,000	1,953,808
85,000	1,934,435

As we move down this table, we find that small increases in the production quantity are estimated to have rather large changes in expected profit that alternate strangely between increases and decreases. Most striking is the result of increasing production from 83,000 to 84,000, which is estimated here to increase expected profit by more than $100,000, even though we cannot earn more than $28 per snowboard! This absurd conclusion is a result our using different simulated demands to evaluate different production quantities. In some simulations, large demands may have occurred more frequently in the simulations when we were evaluating the production quantity 84,000 in cell C17, and less frequently in the simulations when we were evaluating the production quantity 83,000 in cell C17.

The difficulty with this approach is that our alternative strategies are being evaluated with different simulated values of the random variables. So a strategy that is always worse than some other strategies might appear better in such analysis, if the random variables that were used to generate its simulated payoffs happened to be more favorable (i.e., included more high-demand outcomes and fewer low-demand outcomes than the random data that were used to evaluate other strategies). Using larger data sets can reduce the probability of this kind of error, but it may be better to avoid it altogether by using the same simulated random variables to evaluate all of our alternative strategies.

6.3 METHOD 2: SIMULATING PAYOFFS FROM ALTERNATIVE STRATEGIES SIMULTANEOUSLY

We can compute payoffs under several alternative strategies in our simulation model, with the same values of the simulated random variables, and the payoffs from these alternative strategies can be listed in the model output row of the simulation table. Then payoffs from all these strategies will be tabulated at once. So we will get a simulation table that has a payoff series for each strategy in one simulation table, where all these payoff series were computed using the same series of simulated random variables.

To say which production quantity is better under the expected value criterion, we need to estimate how expected profit changes when the production quantity changes. That is, what we really need to estimate accurately are the differences between expected profits under different production quantities. These differences can be estimated are more accurately in problems like Peak Technology when the various production quantities are evaluated with the same series of simulated demands, as illustrated in Figure 6.3.

Method 2 is applied in Figure 6.3 for the selected production quantities of 85,000, 95,000, 105,000, and 115,000, which have been entered into cells B20:E20. The simulated demands to compare these alternative decisions are generated by the random variables in cells B16 and C16 of Figure 6.3.

Cell B16 simulates the weather pattern by the formula

$$= IF(RAND()<B8,1,0),$$

where B8 contains the value $1/3 = P(Cold)$. The simulated demand in cell C16 is generated by the formula

$$= IF(B16=1,EXP(NORMINV(RAND(),LN(C5),(LN(C6)-$$

$$LN(C4))/1.349)),NORMINV(RAND(),B5, (B6-B4)/1.349))$$

where cells C4:C6 contain the demand quartiles for cold weather, and cells B4:B6 contain the demand quartiles for normal weather.

The profit that would result from the demand in cell C16 and the production quantity in cell B19 is computed in cell B29 of Figure 6.3 by the formula

=($B$12-$B$10)*B$20+($B$11-$B$12)*MIN(B$20,$C16)

To interpret this formula, notice that B12 contains the remainder value and B10 contains the cost per unit, so (B12-B10)*B20 is the (negative) profit that would result from producing the quantity B20 and leaving all production in our final inventory. But then taking each unit out of inventory and selling it at the price B11 increases profit by B11- B12, and the number of units sold is the minimum of production and demand MIN(B20,C16). Copying this formula from B29 to B29:E29 gives us the alternative profits that would result from the alternative production quantities in B20:E20, all with the same simulated demand in C16.

The range B30:E430 in Figure 6.3 contains a table of profit outcomes for these four production quantities with 401 independent simulations of the demand random variable B16. The sample averages of these simulated profits are computed in cells B22:E22 to estimate the expected profit for each production quantity. By the criterion of expected profit maximization, the best of these four production quantities is apparently 95,000, which yields an estimated expected profit of $1,968,541. These estimated expected profits are plotted above their corresponding production quantities by the "mean" curve in the chart in Figure 6.3.

The top and bottom curves in the chart in Figure 6.3 show the values at the 0.90 and 0.10 cumulative probability levels in the profit distributions that result from these four alternative production quantities (as computed in cells B24:E25). The increasing spread between these curves shows that higher production quantities may also be associated with more risk.

The difficulty with this method is that it requires us to specify in advance the strategies that we want to evaluate, and the simulation table can become big and unwieldy if we want to evaluate many strategies.

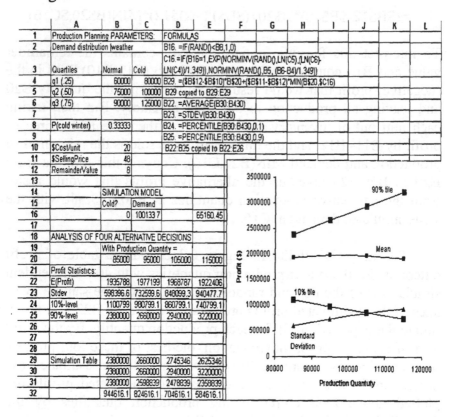

	A	B	C	D	E	F	G	H	I	J	K	L
1	Production Planning PARAMETERS:			FORMULAS								
2	Demand distribution	weather		B16. =IF(RAND()<B8,1,0)								
3	Quartiles	Normal	Cold	C16.=IF(B16=1,EXP(NORMINV(RAND(),LN(C5),(LN(C6)-LN(C4))/1.349)),NORMINV(RAND(),B5, (B6-B4)/1.349))								
4	q1 (.25)	60000	80000	B29. =($B$12-$B$10)*B$20+($B$11-$B$12)*MIN(B$20,$C16)								
5	q2 (.50)	75000	100000	B29 copied to B29:E29								
6	q3 (.75)	90000	125000	B22. =AVERAGE(B30:B430)								
7				B23. =STDEV(B30:B430)								
8	P(cold winter)	0.33333		B24. =PERCENTILE(B30:B430,0.1)								
9				B25. =PERCENTILE(B30:B430,0.9)								
10	$Cost/unit	20		B22:B25 copied to B22:E26								
11	$SellingPrice	48										
12	RemainderValue	8										
13												
14		SIMULATION MODEL										
15		Cold?	Demand									
16		0	100133.7		65160.45							
17												
18	ANALYSIS OF FOUR ALTERNATIVE DECISIONS											
19		With Production Quantity =										
20		85000	95000	105000	115000							
21	Profit Statistics:											
22	E(Profit)	1935788	1977199	1968787	1922406							
23	Stdev	598396.6	732599.6	848099.3	940477.7							
24	10%-level	1100799	980799.1	860799.1	740799.1							
25	90%-level	2380000	2660000	2940000	3220000							
26												
27												
28												
29	Simulation Table	2380000	2660000	2745346	2625346							
30		2380000	2660000	2940000	3220000							
31		2380000	2598839	2478839	2358839							
32		944616.1	824616.1	704616.1	584616.1							

Figure 6.3 Simulated payoffs from alternative snowboard production strategies.

6.4 METHOD 3: OPTIMIZING ENGINEERING DECISION VARIABLES TO MAXIMIZE PAYOFF

If the number of random variables that influence payoffs under all strategies is not too large, then we can make a table of simulated values of these random variables. With this fixed simulation table of all payoff-relevant random variables, we can then compute payoffs for as many different strategies as we like, and the expected payoffs and other summary statistics can be summarized for different decisions in a data table. With this approach, we can even ask Excel Solver to search for strategies that maximize the expected payoff or some other target formula. This approach can become unwieldy and not feasible, however, if the number of random variables that we need to tabulate is very large, in which case in which case we should use approach 2 above instead.

Figure 6.4 applies the third approach to the Peak Technology case: tabulating the random variables first and then computing profits from this fixed simulation data by formulas that depend "live" on a cell that represents the decision variable. The random variables for weather and demand are simulated in cells B22 and C22 of Figure 6.4 by the formulas

=IF(RAND()<B8,1,0)

to simulate the weather pattern (cold=1, normal=0) in cell B22, and

=IF(B22=1,GENLINV(RAND(),C4,C5,C6),GENLINV(RAND(),B4,B4,B6)

to simulate Peak Technology's snowboard demand in cell C22. Fixed data from 1001 simulations of these random variables has been stored below in the range B23:C1023. So the expected value of demand can be estimated to be 85,016, according to the average of sampled demands that is computed in cell C18.

A possible production quantity has been entered into cell D19 of Figure 6.4, and the profits that would result from this production quantity for each of the simulated demand values are computed in E23:E1023 by entering the formula

=($B$12-$B$10)*$D$18+($B$11-$B$12)*MIN($D$18,C23)

into cell E23, and then copying E23 to cells E23:E1023. Then the expected profit with the production quantity D18 is computed in cell E18 by the formula

=AVERAGE(E23:E1023)

The big advantage of this approach is that this estimated expected profit in cell E18 of Figure 6.4 depends actively on the decision variable in cell D18. We can try different decisions in D18 and watch how the expected profits change in E18. A data table can be used to show how expected profit and other statistics of the simulated profit distribution would vary with any group of selected production quantities, as illustrated in G15:K22 of Figure 6.4.

Microsoft Excel comes with an add-in called Solver.xla which can solve optimization problems in spreadsheets. The default installation of Excel often does not install this xla file, and so you may need to redo a custom installation to get Solver.xla into the appropriate library folder in your computer's hard disk. Once it is there, you need to use the Tools>AddIns menu to activate Solver.xla as a regular part of Excel. Once installed and added in, Solver should appear as a part of your Tools menu in Excel.

We can ask Solver to search for the value of our decision variable in cell D18 that would maximize the expected profit in cell E18. The value 98,077 that is shown in cell D18 in Figure 6.4 is actually the result of Solver's optimization, when directed to maximize E18 by changing D18. (The global optimality of this result was confirmed by repeated applications of Solver with different initial conditions, some much higher and others much lower, all of which terminated with this production quantity reported as the optimal solution.) So we may estimate that 98,077 snowboards is Peak Technology's optimal production quantity, according to the criterion of expected profit maximization, and that Peak Technology can achieve an expected profit of about $1,986,461 with this production quantity. (A similar conclusion might have been taken from the analysis in Figure 6.3 with different simulation data, except that there we had no way to check whether higher expected profits could have been achieved by other production quantities between 85,000 and 95,000, or between 95,000 and 105,000.)

	A	B	C	D	E	F	G	H	I	J
1	SNOWBOARDS PRODUCTION PARAMETERS:			FORMULAS FROM RANGE A1:E1023						
2	Demand distribution	weather		B22. =IF(RAND()<B8,1,0)						
3	Quartiles	Normal	Cold	C22. =IF(B22=1,EXP(NORMINV(RAND(),LN(C5),(LN(C6)-LN(C4))/1.349)),NORMINV(RAND(),B5, (B6-B4)/1.349))						
4	q1 (.25)	60000	80000	C18. =AVERAGE(C$23:C$1023)						
5	q2 (.50)	75000	100000	C19. =STDEV(C$23:C$1023)						
6	q3 (.75)	90000	125000	C18:C19 copied to E18:E19						
7				E23. =($B$12-$B$10)*$D$18+($B$11-$B$12)*MIN($D$18,C23)						
8	P(Cold)	0.33333		E23 copied to E23.E1023						
9				SOLVER: maximize E18 by changing D18						
10	$Cost/unit	20								
11	$SellingPrice	48								
12	RemainderVal	8								
13										
14										
15										
16							98077.13	1986461	890897.6	2746160
17			Demand	ProdnQ	Profit statistics		70000	1778584	1227823	1960000
18		Mean	87402.09298	98077.13	1986461	E(Profit)	80000	1903668	1107823	2240000
19		Stdev	32578.54506		753293.3	Stdev	90000	1968660	987823.2	2520000
20							100000	1984931	867823.2	2800000
21		Cold?	Demand				110000	1956912	747823.2	3080000
22	Simulation Table	0	79134.9226		Profit		120000	1903819	627823.2	3360000
23		0	69850.79529		1617106					
24		0	44890.3992		618690.4		FORMULAS FROM RANGE G15:K22			
25		0	138223.1635		2746160		H16. =E18		k16 =E20	
26		1	200405.1638		2746160		I16. =PERCENTILE(E23:E1023,0.1)			
27		0	92454.58712		2521258		J16. =PERCENTILE(E23:E1023,0.9)			
28		0	93535.99419		2564514		H17:J22. {=TABLE(,D18)}			
29		1	98129.09647		2746160					
30		0	48022.54321		743976.1					

Figure 6.4 Optimize snowboards production strategies to maximize payoffs.

Notice that this optimal production quantity for maximizing the expected value of profit is quite different from the expected value of demand which we estimated in cell C18 of Figure 6.4. If demand were perfectly predictable then the profit-maximizing production quantity would be equal to demand. But when there is uncertainty about demand, the optimal production quantity is not necessarily equal to expected demand, even when our goal is to maximize expected profit.

6.5 VALUE OF INFORMATION FOR ENGINEERING DECISIONS

In a production planning session, a marketing manager remarked that good forecasts of the coming winter's general weather would be available in November. But the manufacturing director replied that a delay of snowboard production until November could substantially increase Peak Technology's total productions costs, perhaps by $100,000 or more.

This part of the Peak Technology case raises the question of whether Peak Technology should be willing to pay some additional costs of about $100,000, to be able to base its production decision on better information about the future weather pattern that will prevail next winter.

In effect, our problem is to estimate how much information about the weather pattern would be worth to Peak Technology. Figure 6.5 shows a spreadsheet to calculate the expected value of this information for Peak Technology, using the criterion of expected profit maximization (risk neutrality). Figure 6.5 was constructed from the spreadsheet shown in Figure 6.4, and the tables of simulated weather and demand data are the same in Figures 6.4 and 6.5.

To investigate the question of what would be optimal if Peak Technology knew that the weather pattern would be cold, the value 1 was entered into cell G21 of Figure 6.5. Then to select the profits from the D column in the cases where cold weather happens, the formula

=IF(B23=$G$21,E23,"..")

was entered into cell G23, and G23 was copied in cells G23:G1023. The result is that the G23:G1023 range selects all the profits from the E column in rows where the weather pattern is cold, but the profits are omitted (and their place is taken by the text "..") in rows where the weather pattern is normal. Excel's AVERAGE and STDEV functions ignore non-numerical data. So the formula

=AVERAGE(G23:G1023)

in cell G18 returns our estimate of Peak Technology's conditionally expected profit when it uses the production quantity in cell D18 and the weather pattern is cold. By changing the value of cell D18, it can be shown in this spreadsheet that an order quantity of approximately 118,000 would maximize Peak Technology's conditionally expected profit given cold weather, as computed in cell G18.

Next we can change the value of cell G21 to 0. The result is that the G23:G1023 range instead selects all the profits from the E column in rows where the weather pattern is normal (not cold), and the profits will be omitted in rows where the weather pattern is cold. So with G21 equal to 0, the average reported in cell G18 becomes an estimate of the conditionally expected profit given normal weather, when the production quantity is as in cell D18. Adjusting the value of cell D18 with cell G21 equal to 0, we can find that an order quantity of approximately 87,000 would maximize the conditionally expected profit given normal weather, as computed in cell G18.

The I,J,K columns of the spreadsheet show how to do this analysis in one step, using strategic analysis. Decision analysts use the word strategy to refer to any plan that specifies what decisions would be made as a function of any information that the decision maker may get now and in the future. So suppose that Peak Technology's managers anticipate getting a perfect forecast of the winter weather pattern before they choose the production quantity, but they do not yet have the forecast in hand. In this situation, a strategy for

Peak Technology would be any rule that specifies how much to produce if the weather forecast is normal and how much to produce if the forecast is cold.

The range I18:J19 in Figure 6.5 is used to display the strategy being considered. Cells I18 and I19 contain the numbers 0 and 1 respectively, to represent the cases of normal and cold forecasts. The intended order quantities in these two cases are listed in cells J18 and J19 respectively. For example, entering the value 90,000 into cell J18 and 100000 into cell J19 would represent the strategy of planning to order 90,000 snowboards if the forecast is normal but 100,000 snowboards if the forecast is cold.

	A	B	C	D	E	F	G	H	I	J	K	L
1		SNOWBOARDS PRODUCTION PARAMETERS		FORMULAS FROM RANGE A1:E1023								
2	Demand distribution (weather			B22 =IF(RAND()<B8,1,0)								
3		Quartiles Normal	Cold	C22 =IF(B22=1 EXP(NORMINV(RAND(),LN(C5),(LN(C6)-LN(C4))/1.349);,NORMINV(RAND(),B5, (B6-B4)/1.349))								
4	q1 (25)	60000	80000	E23 =($B$12-$B$10)*$D$18+($B$11-$B$12)*MIN($D$18,C23)								
5	q2 (50)	75000	100000	G23 =IF (B23=$G$21,E23,".")								
6	q3 (75)	90000	125000	J23 =VLOOKUP(B23,$I$18:$J$19,2,0)								
7				K23 =($B$12-$B$10)*J23+($B$11-$B$12)*MIN(J23,C23)								
8	P(Cold)	0.33333		E23:K23 copied to E23:K1023.								
9				E18 =AVERAGE(E23:E1023)			E18 copied to B18					
10	$Cost/unit	20		E19 =STDEV(E23:E1023)								
11	$SellingPrice	48		E18:E19 copied to G18:G19,K18:K19.								
12	RemainderVal	8		K15 =K18-E18								
13				SOLVER: maximize K18 by changing J18:J19.								
14											E(ValueOfInfo)	
15				DECISION ANALYSIS							5363863	
16				Without forecast			Conditional		Strategy with forecast			
17				ProdnQ	Profit statistics		on weather		with	ProdnQ	Profit	
18		0.354645355		98077.13	1986460.881	E(Profit)	2337635.06		0	89077.895	2040100	E(Profit)
19					753293.2744	Stdev	578052.362		1	123312.77	835534.5	Stdev
20							with (Cold?)=					
21		Cold?	Demand				1					
22	Simulation Table	1	140685.4519	Profit	Profit	(Cold?)				ProdnQ	Profit	
23		0	69850.79529	1617106.227		.			89077.895	1725097	
24		0	44890.3992	618690.3831		..			89077.895	726681.2	
25		0	138223.1635	2746159.698		.			89077.895	2494181	
26		1	200405.1638	2746159.698	2746159.7				123312.77	3452757	
27		0	92454.58712	2521257.9		.			89077.895	2494181	
28		0	93535.99419	2564514.183		.			89077.895	2494181	
29		1	98129.09647	2746159.698	2746159.7				123312.77	2445411	
30		0	48022.54321	743976.1436		.			89077.895	851967	

Figure 6.5 Evaluate expected value of information.

The formula

$$=VLOOKUP(B23,\$I\$18:\$J\$19,2,0)$$

has been entered into cell J23 in Figure 6.5. This formula tells Excel to look for the value of B23 in the leftmost (I) column of the range I18:J19, and then return the value found in column 2 within this range (the J column). Including 0 as the optional fourth parameter of the VLOOKUP function tells it to look for an exact match for the B23 value in the I18:I19 range. Then this formula in cell J23 has been copied to J23:J1023. The result is that each cell in J23:J1023 displays the production quantities that would be implemented under our strategy for the weather pattern given in the corresponding cell in B23:B1023. Then the profits that would result from implementing this strategy are computed in the K column by entering the formula

$$=(\$B\$12-\$B\$10)*J23+(\$B\$11-\$B\$12)*MIN(J23,C23)$$

into cell K23 and then copying cell K23 to K23:K1023.

Peak Technology's expected profit under this I18:J19 strategy is computed in cell K18 by the formula

$$=AVERAGE(K23:K1023)$$

Now we can ask Solver to maximize the target cell K18 by changing the strategic plan's production quantities in cells J18:J19. The result is that Solver will find the optimal strategy for using this forecast information to maximize Peak Technology's expected profit. The resulting values returned by Solver are actually shown in cells J18:J19 of Figure 6.5. That is, the optimal strategy found by Solver is to produce 87,226 snowboards if the forecast is normal and to produce 118,443 snowboards if the forecast is cold. In fact, these two conditional production quantities are the same as the quantities that would maximize the conditionally expected profit in G18 given each of these two weather patterns in G21.

Figure 6.5 also shows in cell D18 the optimal production quantity for maximizing the expected profit in cell E18 when Peak Technology does not get any weather forecast, as we saw in Figure 6.4 with the same simulation data. So the difference between cell E18 and cell K18, which is computed in cell K15, is the amount that Peak Technology's expected profit would be increased by allowing Peak Technology to use such a perfect weather forecast in its production planning. This increase in expected profit is called the *expected value of information*. So according to our simulation analysis, the expected value of weather information to Peak Technology would be approximately $45,000. This expected value of information is the most that Peak Technology should be willing to pay to get a perfect forecast of the winter weather pattern before choosing its production quantity. Thus, the estimated costs of $100,000 to delay production until forecast information is available are greater than the expected benefits of getting the information.

6.6 DECISION CRITERIA: HOW TO VALUE ENGINEERING ALTERNATIVES

The objective of engineering decision analysis is to make the "best" decision. Only in rare occasions would an engineer be comparing alternatives whose payoffs are known deterministically. In this case, the engineer would obviously choose the alternative with the highest payoffs. For real-life engineering decision analysis problems involving uncertainty, the "best" decision may mean different things to different people and at different times.

Consider a simple decision problem, in which an engineering firm has an option to install an emergency electrical power system to guard against potential electrical power failure. The additional cost for the system will be $2000 per year; whereas if power failure occurs and no emergency power supply is available, the engineering firm would incur a loss of $10,000. Based on experience and judgment, the engineering firm believes that the annual probability of power failure is only 10%. Assume that the chance of more than one power failure in a year is negligible. The decision tree for this example is shown in Figure 6.6. If installing the emergency system, the engineering firm would have to spend $2000; if not, the engineering firm has a 90% chance of not spending any money but also a 10% chance of losing $10,000.

If extremely pessimistic, the engineering firm would try to minimize the potential loss by installing the emergency system, because the maximum loss in this alternative is only $2000, relative to $10,000 for the other alternative. On the other hand, if extremely optimistic, the engineering firm would not install the emergency power system, since this alternative will yield the maximum gain; namely, it may cost the engineering firm nothing compared to $2000 for the other alternative.

The first basis of selecting the optimal alternative is called the *mini-max* criterion, in which the decision maker selects the alternative that minimizes the maximum loss. In the later case, the engineering firm tries to maximize the maximum possible gain among the alternatives, referred to as the *maxi-max* criterion.

Nether of these two criteria are practical in the long run. If the mini-max decision maker consistently follows this criterion, he will never venture into a decision that may result in substantial gain as long as there is a finite (though small) chance that he may be incurring a loss, whereas the maxi-max decision maker will never choose the realistic alternative provided that there is a finite chance for a more attractive outcome. Any rational decision maker should consider the relative likelihood associated with the gain or loss in each alternative, rather than strictly following these criteria.

When the consequences associated with each alternative in a decision analysis can be expressed in terms of monetary values, a widely used criterion is the *maximum expected monetary value*, as we discussed in Chapter 3.

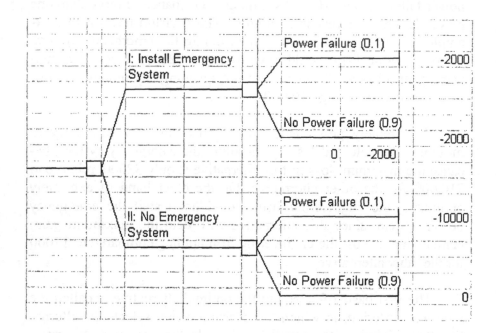

Figure 6.6 Decision tree for installation of emergency power system.

Example 6.1 Suppose the engineering firm referred above follows the maximum expected monetary value decision criterion, the firm would compute the expected monetary value of the two alternatives as:

$$E(\text{Installing}) = 0.1 \times (-2000) + 0.9 \times (-2000)$$
$$= -2000 \text{ dollars};$$

and

$$E(\text{Not Installing}) = 0.1 \times (-10,000) + 0.9 \times (0)$$
$$= -1000 \text{ dollars}$$

Both the maxi-max and mini-max criteria ignore the probabilities of the possible outcomes of an alternative. The available information is, therefore, not fully utilized in the decision analysis. With the maximum expected monetary criterion, a decision maker is systematically weighting the value of each outcome by the corresponding probability.

It is conceivable that an event with an occurrence probability of 99.99% may still fail to occur, in spite of the extreme odd. Nevertheless, if a decision maker consistently bases his decisions on the maximum expected monetary criterion, the total monetary value obtained from all his decisions (in the long run) will be maximum.

In the decision analysis literature, a decision maker is called risk neutral if he (or she) is willing to base his decisions purely on the criterion of maximizing the expected value of his monetary income. The criterion of maximizing expected monetary value is so simple to work with that it is often used as a convenient guide to decision making even by people who are not perfectly risk neutral. But in some situations, people often feel that comparing gambles only on the basis of expected monetary values would take insufficient account of their aversion to risks.

For example, suppose an engineer is considering an innovative technology with $20,000 potential payoff. The chance of success for this innovative technology is 50%. If the engineer is risk neutral, then she or he should be unwilling to adopt any other engineering alternatives with payoff less than the expected payoff value for the innovative technology, which is $10,000. But many risk averse engineers might decide a proven technology which can assure a certain payoff of $9000.

Given any such uncertain alternative that promises to have a payoff that will be drawn randomly from some probability distribution, a decision maker's certainty equivalent of this uncertain alternative is the lowest amount of payoff-for-certain that the decision maker would be willing to accept instead of this uncertain alternative. That is, if $7000 would be your certainty equivalent of this uncertain alternative that would pay you either $20,000 or $0, each with probability 1/2, then you should be just indifferent with probability 1/2, then you should be just indifferent between this alternative or another alternative which assures a payoff of $7000.

In these terms, a risk-neutral decision maker is one whose certainty equivalent of any alternative is just equal to its expected monetary value (EMV). A decision maker is risk averse if his or her certainty equivalent of project is less than the alternative's expected monetary value. The difference between the expected monetary value of an alternative and a risk-averse decision maker's certainty equivalent of the alternative is called the decision maker's risk premium for the alternative. (RP=EMV-CE.)

When you have a choice among different engineering alternatives, you should choose the one for which you have the highest certainty equivalent, because it is the one that you would value the most highly. But when an alternative is complicated, you may find it difficult to assess your certainty equivalent for it. The great appeal of the risk-neutrality assumption is that, by identifying your certainty equivalent with the expected monetary value, it makes your certainty equivalent something that is straightforward to compute or estimate by simulation. So what we need now is to find more general formulas that risk-averse decision makers can use to compute their certainty equivalents for complex engineering alternatives and monetary risks.

A realistic way of calculating certainty equivalents must include some way of taking account of a decision maker's personal willingness to take risks. The full diversity of formulas that a rational decision maker might use to calculate certainty equivalents is described by a branch of economics called utility theory. Utility theory generalizes the principle of expected value maximization in a simple but very versatile way. Instead of assuming that people want to maximize their expected monetary values, utility theory instead assumes that each individual has a personal utility function that assigns a utility values to all possible monetary income levels that the individual might receive, such that the individual always wants to maximize his or her expected utility value. So according to utility theory, if the random variable **X** denotes the amount of money that some uncertain alternative would pay you, and if U(y) denotes your utility for any amount of money y, then your certainty equivalent CE of this uncertain alternative should be the amount of money that gives you the same utility as the expected utility of the alternative. That is, we have the basic equation

$$U(CE) = E(U(X)).$$

Utility theory can account for risk aversion, but it also is consistent with risk neutrality or even risk-seeking behavior, depending on the shape of the utility function. In 1947, von Neumann and Morgenstern gave an ingenious argument to show that any consistent rational decision maker should choose among alternatives according to utility theory. Since then, decision analysts have developed techniques to assess individuals' utility functions. Such assessment can be difficult, because people have difficulty thinking about decisions under uncertainty, and because there as so many possible utility functions. But just as we could simplify subjective probability assessment by assuming that an individual's subjective probability distribution for some unknown quantity might be in some natural mathematical family of probability distributions, so we can simplify the process of subjective utility assessment by assuming that a decision maker's utility function in some natural family of utility functions.

For practical decision analysis, the most convenient utility functions to use are those that have a special property called *constant risk tolerance*. Constant risk tolerance means that if we change an alternative by adding a fixed additional amount of payoff, about which there is no uncertainty, then the certainty equivalent should increase by the same amount. This assumption of constant risk tolerance gives us a simple one-parameter family of utility functions: If a decision maker is risk averse and has constant risk tolerance, then the decision maker's preferences can be described by a utility function of the

$$U(x) = \text{-EXP(-x/r)}$$

where r is a number that is the decision maker's risk tolerance constant.

Example 6.2 Consider the problem discussed in Example 6.1 in which the engineering firm is deciding on whether or not to install an emergency power system. Suppose the risk tolerance constant is $5000, thus the utility function for money is

$$U(x) = e^{-x/5000} \qquad (x<0)$$

The expected utility value of the first alternative (for installation) is thus

$$E(U_1) = 0.1(-e^{2000/5000}) + 0.9(-e^{2000/5000}) = -1.49$$

Similarly, for the no-installation alternative

$$E(U_2) = 0.1(-e^{1000/5000}) + 0.9(-e^{1000/5000}) = -1.64$$

According to the maximum expected utility value function, the engineering firm should install the emergency power supply system. Then the engineering firm can avoid the potential risk of losing $10,000, which is a considerable risk to the firm given the risk tolerance constant of $5000. Although installing the system would result in maximum expected monetary value (see Example 6.1), the engineering firm's risk-aversive behavior as described by the utility function causing the firm to select the more conservative alternative.

This assumption of constant risk tolerance is very convenient in practical decision analysis, because it allows us to evaluate independent alternatives separately. If you have constant risk tolerance then, when you are going to earn money from two independent alternatives, your certainty equivalent for the sum of the two independent alternatives is just the sum of your certainty equivalent for each alternative by itself. That is, your certainty equivalent value of an alternative is not affected by having other independent alternatives, provided that you have one of these constant-risk-tolerance utility functions.

6.7 EVALUATION OF CASH FLOW INFORMATION: PAY- BACK METHOD

The payback method of analysis evaluates projects based on how long it takes to recover the amount of money put into the project. The shorter the payback period, the better. Certainly there is some intuitive appeal to this method. The sooner we get our money out of the project, the lower the risk. If we have to wait a number of years for a project to "pay off," all kinds of things can go wrong. Furthermore, given high interest rates, the longer we have our initial investment tied up, the more costly it is for us.

EXHIBIT 1

Payback Method--Alternative Projects

Project Cash Flows

	One	Two	Three	Four
January 1992	$(400)	$(400)	$(400)	$(400)
1992	0	0	399	300
1993	1	399	0	99
1994	399	1	0	0
1995	$ 500	$ 500	$ 500	$5,000
TOTAL	$ 500	$500	$499	$4,999

Exhibit 1 presents an example for the payback method. Although there may be intuitive support for this method, we will note a number of weaknesses. In the exhibit, four alternative projects are being compared. In each project, the initial outlay is $400. By the end of 1994, projects one and two have recovered the initial $400 investment. Therefore, they have a payback period of three years. Projects three and four do not recover the initial investment until the end of 1995. Their payback period is four years, and they are therefore considered to be inferior to the other two projects.

It is not difficult at this point to see one of the principal weaknesses of the payback method. It ignores what happens after the payback period. The total cash flow for project four is much greater than the cash received from any of the other projects, yet it is considered to be one of the worst of the projects. In a situation where cash flows extend for 20 or 30 years, this problem might not be as obvious, but it could cause us to choose incorrectly.

Is that the only problem with this method? No. Another obvious problem stems from the fact that according to this method, projects one and two are equally attractive because they both have a three-year payback period. Although their total cash flows are the same, the timing is different. Project one provides one dollar in 1993, and then $399 during 1994. Project two generates $399 in 1993 and only $1 in 1994. Are these two projects equally as good because their total cash flows are the same? No. The extra $398 received in 1993 from project two is available for investment in other profitable opportunities for one extra year, as compared to project one. Therefore, it is clearly superior to project one. The problem is that the payback method doesn't formally take into account the time value of money.

This deficiency is obvious in looking at project three as well. Project three appears to be less valuable than projects one or two on two counts. First, its payback is four years rather than three, and second, its total cash flow is less than either project one or two. But if we consider the time value of money, then project three is better than either project one or two. With project three, we get the $399 right away. The earnings on that $399 during 1993 and 1994 will more than offset the shorter payback and larger cash flow of projects one and two.

Although payback is commonly used for a quick and dirty project evaluation, problems associated with the payback period are quite serious. As a result, there are several methods commonly referred to as discounted cash flow models that overcome these problems. Later, we will discuss the most commonly used of these methods, net present value and internal rate of return. However, before we discuss them, we need to specifically consider the issues and mechanics surrounding time value of money calculations.

6.8 THE TIME VALUE OF MONEY

It is very easy to think of projects in terms of total dollars of cash received. Unfortunately, this tends to be fairly misleading. Consider a project in which we invest $400 and in return we receive $520 after three years. We have made a cash profit of $120. Because the profit was earned over a three-year period, it is a profit of $40 per year. Because $40 is 10% of the initial $400 investment, we have apparently earned a 10% return on our money. While this is true, that 10% is calculated based on simple interest.

Consider putting money into a bank that pays a 10% return "compounded annually." The term compounded annually means that the bank calculates interest at the end of each year and adds the interest onto the initial amount deposited. In future years, interest is earned not only on the initial deposit, but also on interest earned in prior years. If we put $400 in the bank at 10% compounded annually, we would earn $40 of interest in the first year. At the beginning of the second year we would have $440. The interest on the $440 would be $44. At the beginning of the third year, we would have $484 (the $400 initial deposit plus the $40 interest from the first year, plus the $44 interest from the second year). The interest for the third year would be $48.40. We would have a total of $532.40 at the end of three years.

The 10% compounded annually gives a different result from the 10% simple interest. We have $532.40 instead of $520 from the project. The reason for this difference is that in the case of the project, we did not get any cash flow until the end of the project. In the case of the bank, we were essentially given a cash flow at the end of each year. We reinvested that cash flow in the bank, although we could have withdrawn the interest from the bank and invested it elsewhere. The crucial fact is that the total amount we wind up with is different in these two cases.

The implication of this difference is that we cannot compare two projects that pay the same total return and surmise that they are equal. Two projects requiring investments of $400 are not equally as good if one pays $520 in three years while the other pays $40 after the first year and $40 after the second year and $440 after the third. In both cases the total is $520, but in the latter case $40 is available for investment after the first year and another $40 after the second. We must determine not only how much cash will be received from each project, but also when it will be received. The project that provides the cash flow sooner gives us an opportunity to earn additional profits. What we need is a method that can compare cash flows coming at different points in time and consider the implications of when the cash is received.

Consider a cash amount of $100 today. We refer to it as a present value (PV or P). How much could this cash amount accumulate to if we invested it at an interest rate (i) or rate of return (r) of 10% for a period of time (N) equal to two years? Assuming that we compound annually, the $100 would earn $10 in the first year (10% of $100). This $10 would be added to the $100. In the second year our $110 would earn $11 (that is, 10% of $110). The future value (FV or F) is $121. That is, two years in the future we would have $121.

Mechanically this is a simple process--multiply the interest rate times the initial investment to find the interest for the first period.

Add the interest to the initial investment. Then multiply the interest rate times the initial investment plus all interest already accumulated to find the interest for the second year.

While this is not complicated, it can be rather tedious. Suppose we invest money for 30 years in an investment that compounds monthly. At the end of each month, we calculate interest and add it to the investment balance. For a period of 30 years we have to make this computation 360 times (monthly compounding requires 12 calculations per year for each of the 30 years, or a total of 360 computations).

To simplify this process, mathematical formulas have been developed to solve a variety of "time value of money" problems. The most basic of these formulas states that:

$$FV=PV(1+i)^\wedge N$$

We will not derive this formula here. Many years ago the time value of money calculations could not be performed without an ability to derive, memorize, or look up the formulas. Even once you had the formulas and plugged in the appropriate variables, the mathematical calculation was still somewhat arduous. Later, tables were developed that made the process somewhat easier. A financial manager should have a good understanding of both the formulas and the tables. However, even the tables are somewhat cumbersome and limited.

Modern technology has made our calculations substantially simpler. For an engineering professional, project analysis can be done with the aid of a hand held business-oriented calculator or a spreadsheet. The formulas are built right into the memory of the calculator or as internal functions built into the spreadsheet. If we supply the appropriate raw data, the calculator or spreadsheet performs all of the necessary interest computations.

Traditionalists will be disappointed to discover that this book contains only simple tables and no derivations of the formulas. However, tables and formulas really aren't necessary for either understanding what we are doing, or for computing the results. Calculators, spreadsheets, tables, and formulas all require the same level of understanding from the users--an ability to determine what information is available and what information is being sought.

For instance, if we wanted to know what $100 would grow to in two years at 10%, we would simply tell our calculator or spreadsheet that the present value, P or PV (depending on the brand of calculator or spreadsheet you use) equals $100; the interest rate, %i or i or r, equals 10%; and the number of periods, N, equals 2. Then we would ask the calculator to compute F or FV, the future value.

Can we use this method if compounding occurs more frequently than once a year? Bonds often pay interest twice a year. Banks often compound quarterly on savings accounts. On the other hand, banks often compound monthly to calculate mortgage payments. Using our example of $100 invested for two years at 10%, we could easily adjust the calculation for semi-annual, quarterly, or monthly compounding. For example, for semi-annual compounding, N becomes 4 because there are two semi-annual periods per year for two years. The rate of return, or interest rate, becomes 5%. If the rate earned is 10% for a full year, then it is half of that, or 5% for each half year.

For quarterly compounding, N equals 8 (four quarters per year for two years) and i equals 21/2 percent (10% per year divided by four quarters per year). For monthly compounding, N equals 24 and i equals 10%/12. Thus, for monthly compounding, we would tell the calculator that PV=$100, i=10%/12, and N=24. Then we would tell the calculator to compute FV. We need a calculator designed to perform present value functions in order to do this.

If we expect to receive $121 in two years can we calculate how much that is worth today? This question calls for a reversal of the compounding process. Suppose we would normally expect to earn a return on our money of 10%. What we are really asking here is, "How much would we have to invest today at 10%, to get $121 in two years?" The answer requires unraveling compound interest. If we calculate how much of the $121 to be received in two years is simply interest earned on our original investment, then we know the present value of that $121. This process of removing or unraveling the interest is called discounting. The 10% rate is referred to as a discount rate. Using a calculator or spreadsheet, this is a simple process. We again supply the i and the N, but instead of telling the calculator the PV and asking for the FV, we tell it the FV and ask it to calculate the PV.

Let's consider a problem of whether to accept $250 today, or $330 in 22 months. Assume that we can invest money in a project with a 10% return and monthly compounding. Which choice is better? We can tell our calculator that FV=$330, N=22, and i=10%/12. If we then ask it to compute PV, we find that the present value is $275. This means that if we invest $275 today at 10% compounded monthly for 22 months, it accumulates to $330. That is, receiving $330 in 22 months is equivalent to having $275 today. Because this amount is greater than $250, our preference is to wait for the money, assuming there is no risk of default. Looking at this problem another way, how much would our $250 grow to if we invested it for 22 months at 10%? Here we have PV=$250, N=22, and i=10%/12. Our calculation indicates that the FV=$300. If we wait, we have $330 22 months from now. If we take $250 today and invest it at 10%, we only have $300 22 months from now. We find that we are better off to wait for the $330, assuming we are sure that we will receive it.

Are we limited to solving for only the present or future value? No, this methodology is quite flexible. Assume, for example, that we wish to put $100,000 aside today to pay off a $1,000,000 loan in 15 years. What rate of return must be earned, compounded annually, for our $100,000 to grow to $1,000,000? Here we have the present value, or $100,000, the number of periods, 15 years, and the future value, or $1,000,000. It is a simple process to determine the required rate of return. If we simply supply our calculator or spreadsheet with the PV, FV, and N, the calculator or spreadsheet readily supplies the i, which is 16.6% in this case.

Or, for that matter, if we had $100,000 today and knew that we could earn a 13% return, we would calculate how long it would take to accumulate $1,000,000. Here we know PV, FV, and i, and we wish to find N. In this case, N=18.8 years. Given any three of our four basic components, PV, FV, N and i, we can solve for the fourth. This is because the calculator or spreadsheet is simply using our basic formula stated earlier and solving for the missing variable.

So, far, however, we have considered only one single payment. Suppose that we don't have $100,000 today, but we are willing to put $10,000 aside every year for 15 years. If we earn 12%, will we have enough to repay $1,000,000 at the end of the 15 years? There are two ways to solve this problem. We can determine the future value, 15 years from now, of each of the individual payments. We would have to do 15 separate calculations because each succeeding payment earns interest for one year less. We would then have to sum the future value of each of the payments. This is rather tedious. A second way to solve this problem is using a formula that accumulates the payments for us. The formula is:

$$FV = PMT \left[((1+i)N - 1)/i \right]$$

In this formula, PMT represents the payment made each period, or annuity payment. Although you may think of annuities as payments made once a year, an annuity simply means payments that are exactly the same in amount, and are made at equally spaced intervals of time, such as monthly, quarterly, or annually. For example, monthly mortgage payments of $321.48 per month represent an annuity.

To solve problems with a series of identical payments, we have five variables instead of the previous four. We now have FV, PV, N, i, and PMT. However, PV doesn't appear in our formula. There is a separate formula that relates present value to a series of payments. This formula is:

$$PV = PMT[\ 1 - [1/(1+i)N]\ /\ i]$$

Annuity formulas are built into business calculators and spreadsheets. With the calculator or spreadsheet, you can easily solve for PV or i or N or PMT, if you have the other three variables. Similarly, you can solve for FV or i or N or PMT given the other three. For instance, how much would we pay monthly on a 20-year mortgage at 12% if we borrowed $50,000? The present value (PV) is $50,000, the interest rate (%i) is 1% per month, the number of months (N) is 240. Given these three factors, we can solve for the annuity payment (PMT). It is $551 per month.

Annuity formulas provide you with a basic framework for solving many problems concerning receipt or payment of cash in different time periods. Keep in mind that the annuity method can only be used if the amount of the payment is the same each period. If that isn't the case, each payment must be evaluated separately.

6.9 EVALUATION OF CASH FLOW INFORMATION: NET PRESENT VALUE (NPV) METHOD

The net present value (NPV) method of analysis determines whether a project earns more or less than a stated desired rate of return. The starting point of the analysis is determin8ation of this rate.

The Hurdle Rate

The rate of return required in order for a project to be acceptable is called the required rate of return or the hurdle rate. An acceptable project should be able to hurdle over, that is, be higher than this rate.

The rate must take into account two key factors. First, a base profit to compensate for investing money in the project. We have a variety of opportunities in which we could use our money. We need to be paid a profit for foregoing the use of our money in some alternative venture. The second element concerns risk. Any time we enter a new investment, there is an element of risk. Perhaps the project won't work out exactly as expected. The project may turn out to be a failure. We have to be paid for being willing to undertake these risks. The two elements taken together determine our hurdle rate.

The hurdle rate for project analysis differs for different companies. There is no one unique standard required rate of return that can be used by all companies. Different industries tend to have different base rates of return. Further, within an industry one firm may have some advantage over other firms (for example, economies of scale) that allow it to have a higher base return. On top of that, different firms, and even different projects for one firm, have different types and degrees of risk. Buying a machine to use for putting Coca-Cola in bottles involves much less risk than developing new soft drink products.

Historically, U.S. government obligations, particularly treasury bills, have provided a benchmark for rates of return. These short-term government securities provide both safety and liquidity for the investor. Adjusted for inflation, the rate of return on treasury bills has historically been around two to three percent. A firm must be able to do better than this in order to attract any investor's money. This is true even if there is no risk in investing in the firm. We can therefore consider two to three percent to be a lower bound or a base, or "pure" rate of return. By pure rate of return we mean before we consider any risk. Because most firms are more risky than the government, they must also pay their investors for being willing to take the chance that something may go wrong and that they may lose money rather than make a profit. This extra payment is considered to be a return for having taken risks.

One of the most prevalent risks that investors take is loss of purchasing power. That is, price level inflation makes our money less valuable over time. Suppose we could buy a TV for $100, but instead we invest that money in a business. If the firm uses that money to generate a pretax profit of $10 in a year when the inflation rate is 10%, did we have a good year, bad year, or neutral year? Because we have to pay some taxes to the government on our $10 pretax profit, we had a bad year. After paying taxes we have less than $110 at the end of the year. But, due to inflation, it costs $110 to buy a TV set. This means that in deciding if a project is worthwhile, we have to consider whether the rate of return is high enough to cover our after-tax loss of purchasing power due to inflation.

We must also consider a variety of business risks. What if no one buys the product, or they buy it, but fail to pay us? If the product is made or sold internationally, we incur foreign exchange risk and foreign political risk. The specific types of risks faced by a company depend on its industry. The company's past experience with projects like the one being evaluated should be a guide in determining the risk portion of the hurdle rate. For example, if you found that historically one out of every 21 projects you invest in is a complete failure, then you should add 5% to your required rate of return. In that way, the total of the 5% extra profit you earn on each of the 20 successful projects exactly equals the 100% you lose on the unsuccessful project. Some firms build more than 5% into that calculation--they don't want to just break even on the risks they take. They feel they should make extra profits to pay for having been willing to take risks.

When we add the desired base or pure profit return to all of the risk factors, the total is the firm's hurdle rate. In most firms, the top financial officers determine an appropriate hurdle rate or rates and inform non-financial managers that this hurdle rate must be anticipated for a project to receive approval. Therefore, you will not usually have to go through a complete calculation of the hurdle rate yourself.

NPV Calculations

Once we know our hurdle rate, we can use the NPV method to assess whether a project is acceptable. The NPV method compares the present value of a project's cash inflows to the present value of its cash outflows. If the present value of the inflows is greater than the outflows, then the project is profitable because it is earning a rate of return that is greater than the hurdle rate.

For example, suppose that a potential project for our firm requires an initial cash outlay of $10,000. We expect the project to produce a net after-tax cash flow (cash inflows less cash outflows) of $6,500 in each of the two years of the project's life. Suppose our after-tax hurdle rate is 18 %. Is this project worthwhile?

The cash receipts total $13,000, which is a profit of $3,000 overall, or $1,500 per year on our $10,000 investment. Is that a compounded return of at least 18%? At first glance it would appear that the answer is ``no" because $1,500 is only 15% of $10,000. However, we haven't left our full $10,000 invested for the full two years. Our positive cash flow at the end of the first year is $6,500. We are not only making $1,500 profit, but we are also getting back half ($5,000) of our original investment. During the second year, we earn $1,500 profit on a remaining investment of only $5,000. It is not simply how much money you get from the project, but when you get it that is important.

The present value of an annuity of $6,500 per year for two years at 18% is $10,177 (PV=?; PMT=$6,500; N=2; i=18%). The present value of the initial $10,000 outflow is simply $10,000 because it is paid at the start of the project. The NPV is the present value of the inflows, $10,177 less the present value of the outflows, $10,000, which is $177. This number is greater than zero, so the project does indeed yield a return greater than 18%, on an annually compounded basis.

It may not be intuitively clear why this method works, or indeed, that it works at all. However, consider making a deal with your friend who is a banker. You agree that you will put a sum of money into the bank. At the end of the first year, the banker adds 18% interest to your account and you then withdraw $6,500. At the end of year two, the banker credits interest to the balance in your account at an 18% rate. You then withdraw $6,500, which is exactly the total in the account at that time. The account will then have a zero balance. You ask your friend how much you must deposit today in order to be able to make the two withdrawals. He replies, "$10,177."

If we deposit $10,177, at an 18% rate, it will earn $1,832 during the first year. This leaves a balance of $12,009 in the account. We then withdraw $6,500, leaving a balance of $5,509 for the start of the second year. During the second year, $5,509 earns interest of $991 at a rate of 18%. This means that the balance in the account is $6,500 at the end of the second year. We then withdraw that amount.

The point of this bank deposit example is that when we earlier solved for the present value of the two $6,500 inflows using a hurdle rate of 18%, we found it to be $10,177. We were finding exactly the amount of money we would have to pay today to get two payments of $6,500 if we were to earn exactly 18%. If we can invest a smaller amount than $10,177, but still get $6,500 per year for each of the two years, we must be earning more than 18% because we are putting in less than would be needed to earn 18%, but are getting just as much out. Here, we invest $10,000, which is less than $10,177, so we are earning a rate of return greater than 18%.

Conversely, if the banker had told us to invest less than $10,000 (that is, if the present value of the two payments of $6,500 each at 18% was less than $10,000), then it means that by paying $10,000 we were putting in more money than we would have to in order to earn 18%, therefore we must be earning less than 18%.

The NPV method gets around the problems of the payback method. It considers the full project life, and considers the time value of money. Clearly, however, you can see that it is more difficult than the payback method. Another problem with it is that you must determine the hurdle rate before you can do any project analysis. The next method we will look at eliminates the need to have a hurdle rate before performing the analysis.

6.10   EVALUATION OF CASH FLOW INFORMATION: INTERNAL RATE OF RETURN (IRR) METHOD

One of the objections of the NPV method is that it never indicates what rate of return a project is earning. We simply find out whether it is earning more or less than a specified hurdle rate. This creates problems when comparing projects, all of which have positive net present values.

One conclusion that can be drawn from our net present value (NPV) discussion is that when the NPV is zero, we are earning exactly the desired hurdle rate. If the NPV is greater than zero, we are earning more than our required rate of return. If the NPV is less than zero, then we are earning less than our required rate of return. Therefore, if we want to determine the exact rate that a project earns, all we need to do is to set the NPV equal to zero. Because the NPV is the present value (PV) of the inflows less the present value of the outflows, or:

NPV=[PV inflows - PV outflows]

then when we set the NPV equal to zero,

0=[PV inflows - PV outflows]

which is equivalent to: PV inflows=PV outflows.

All we have to do to find the rate of return that the project actually earns, or the "internal rate of return," (IRR) is to find the interest rate at which this equation is true.

For example, consider our NPV project discussed earlier that requires a cash outlay of $10,000 and produces a net cash inflow of $6,500 per year for two years. The present value of the outflow is simply the $10,000 (PV=$10,000) we pay today. The inflows represent a two-year (N=2) annuity of $6,500 (PMT=$6,500) per year. By supplying our calculator or spreadsheet with the PV, N, and PMT, we can simply find the i or r (IRR). In this case, we find that the IRR is 19.4%.

6.11 VARIABLE CASH FLOW

This calculation is simple for any business calculator or spreadsheet that can handle time value of money, frequently called discounted cash flow (DCF) analysis. However, this problem was somewhat simplistic because it assumed that we would receive exactly the same cash inflow each year. In most capital budgeting problems, it is much more likely that the cash inflows from a project will change each year. For example, assume that our $10,000 investment yields $6,000 after the first year and $7,000 at the end of the second year. The total receipts are still $13,000, but the timing of the receipts has changed. The return is not as high because we are receiving $500 less in the first year than we had been in the earlier situation (that is, we get $6,000 instead of $6,500). Although we get $500 more in the second year ($7,000 instead of $6,500), that $500 difference could have been profitably invested if we had it for that intervening year. Are we still over our hurdle rate? What is the current IRR?

From a NPV point of view, it is still rather simple for us to determine whether this reviewed project is still acceptable. We need only find the present value of $6,000 evaluated with an N equal to one year, using our 18% hurdle rate, and add that to the present value of $7,000 evaluated with an N equal to two years at 18%. The sum of these two calculations is the present value of the inflows. We compare this calculation to the $10,000 outflow to see if the NPV is positive. In this case, the present value of the first inflow is $5,085 and the present value of the second inflow is $5,027. Their sum is $10,112, which is still greater than $10,000, so the project is still acceptable.

But what is the project's internal rate of return? The NPV solution was arrived at by two separate calculations. But the IRR method evaluates the one unique rate of return for the entire project. From a mathematical perspective, we run into difficulties if the cash inflows are not the same each year. This creates a situation in which we must solve for the missing i or r by trial and error! There is no better way to do this than to keep trying different rates until we find one rate at which the present value of the inflows is exactly equal to the present value of the outflows. Fortunately, the more advanced of the business calculators and spreadsheets can do this trial and error process for you. However, in purchasing a calculator, you should be sure that it can handle internal rate of return with uneven or variable cash flow. Such a calculator can handle the complexity of having cash receipts and expenditures that differ in amount from year to year over the project's life. All modern spreadsheets including Lotus and Excel handle this problem easily.

In fact, as you watch such a calculator or spreadsheet (on a slow microcomputer) solving this type of problem, you will see numbers flashing across the calculator display for several moments or the spreadsheet pause while recalculating, before the answer finally appears. This is a result of the calculator or computer trying different interest rates, looking for the correct answer. It uses trial and error, just as we would do by hand. Calculating devices aren't smarter than we are, just faster.

6.12 PROJECT RANKING

The NPV method is quite adequate to determine if a project is acceptable or not. Often, however, we may be faced with a situation in which there are more acceptable projects than the number that we can afford to finance. In that case, we wish to choose the best projects. A simple way to do this is to determine the internal rate of return on each project and then to rank the projects from the project with the highest IRR to the lowest. We then simply start by accepting projects with the highest IRR and go down the list until we either run out of money, or reach our minimum acceptable rate of return.

In general, this approach allows the firm to optimize its overall rate of return. However, it is possible for this approach to have an undesired result. Suppose that one of our very highest yielding projects is a parking lot. For a total investment of $50,000, we expect to earn a return of $20,000 a year for the next 40 years. The internal rate of return on that project is 40%. Alternatively, we can build an office building on the same site. For an investment of $10,000,000 we expect to earn $3,000,000 a year for 40 years, or an IRR of 30%. We can only use the site for the parking lot or the building but not both. Our other projects have an IRR of 20%.

If we build the parking lot because of its high IRR, and therefore bypass the building, we will wind up investing $50,000 at 40% and $9,950,000 at 20% instead of $10,000,000 at 30%. This is not an optimal result. We would be better off to bypass the high-yielding parking lot and invest the entire $10,000,000 at 30%.

This contrary outcome is not a major problem. Assuming that we have ranked our projects, if we don't have to skip over any projects because they conflict with a higher-yielding project that we have already accepted, the problem doesn't arise. If we have to skip a small project because we accepted a large project, we still don't have a problem. It is only in the case that we would be forced to skip over a large project (such as a building) because we accepted a small project (such as a parking lot) that causes us concern. In that case, our decision should be based on calculating our weighted average IRR for all projects that we accept. For instance, suppose we have $15,000,000 available to invest. We can invest $50,000 in a parking lot at 40% and $14,950,000 in other projects at 20%, for a weighted average internal rate of return of 20.1%, or we can invest $10,000,000 at 30% and $5,000,000 at 20% for a weighted average IRR of 26.7%. Here we decide to accept the $10,000,000 building at 30%, skipping the parking lot. Although the parking lot itself has a higher IRR, the firm as a whole has a higher IRR for all projects combined if it uses the land for a building.

6.13 SUMMARY

A decision variable is any quantity that we have to choose in a decision problem. To analyze an engineering decision problem, we may begin by constructing a simulation model that describes how profit returns and other outcomes of interest may depend on the decision variables that we control and on the random variables that we do not control. In this model, we can try specifying different values for the decision variables, and we can generate simulation tables that show how the probability distribution of profit outcomes could depend on the values of the decision variables. Using the criterion of expected profit maximization or some other optimality criterion, we can then try to find values of the decision variables that yield the best possible outcome distribution, where "best" may be defined by some optimality criterion like expected value maximization.

The limitations of the expected value maximization as a criterion for decision making have been addressed by an important theory of economic decision making: utility theory. Utility theory shows how to apply an expected value criterion to any individual decision maker who satisfies some intuitive consistency assumptions. The main result of utility theory is that any rational decision maker should have some way of measuring payoff, called a utility function, such that the decision maker will always prefer the alternative that maximizes his or her expected utility. So if a risk-averse decision maker does not want to make all decisions purely on the basis of his expected monetary payoff, that only means that his utility payoff is different from his monetary payoff. Utility theory then teaches us how to assess an individual's risk tolerance and derive a utility function that gives us a utility value corresponding to any monetary value of profit or income. So the good news is that, once we learn how to assess utility functions for risk-averse decision makers, the mathematical method that we will use to analyze any decision problem will still be expected value maximization. The only difference is that, instead of maximizing an expected monetary value, we will maximize an expected utility value. The next chapter will extend the utility functions to engineering decisions balancing multiple objectives.

To adequately evaluate projects, discounted cash flow techniques should be employed. The two most common of these methods are NPV and IRR. The essential ingredient of both of these methods is that they consider the time value of money. A project engineer or engineering manager doesn't necessarily have to be able to compute present values. It is vital, however, that all managers understand that when money is received or paid can have just as dramatic an impact on the firm as the amount received or paid.

REFERENCES

Ahuja, H. N., Dozzi, S. P., Abourizk, S. M. (1994), Project Management, Second Edition, John Wiley & Sons, Inc., New York.

Ang, A. H-S., Tang, W. H. (1984), Probability Concepts in Engineering Planning and Design, Volume II – Decision, Risk, and Reliability," John Wiley & Sons, New York.

Bryant, M. W. (1992), "Risk Management Roundtable: Improving Performance with Process Analysis," Risk Management, 39 #11 (November), pages 47-53.

Burlando, T. (1994), "Chaos and Risk Management," Risk Management, 41 #4 (April), pages 54- 61.

Chicken, J. C. (1994), Managing Risks and Decisions in Major Projects, Chapman & Hall, London, U.K.

Cooper, D. F. (1987), Risk Analysis for Large Projects: Models, Methods, and Cases, Wiley, New York, N.Y.

Defense Systems Management College, (1983), Risk Assessment Techniques: A Handbook for Program Management Personnel, Ft. Belvoir: DSMC.

Englehart, J. P. (1994), "A Historical Look at Risk Management," Risk Management, 41 #3 (March), pages 65-71.

Esenberg, R. W. (1992), "Risk Management in the Public Sector," Risk Management, 39 #3 (March), pages 72-78.

Grose, V. L. (1987), Managing Risk: Systematic Loss Prevention for Executives, Prentice-Hall, Englewood Cliffs, NJ.

Lewis, H.W. (1990), Technological Risk, Norton, New York, N.Y.

Lundgren, R. (1994), Risk Communication: A Handbook for Communicating Environmental, Safety and Health Risks, Battelle Press, Columbus, Ohio.

Kurland, O. M. (1993), "The New Frontier of Aerospace Risks," Risk Management. 40 #1 (January), pages 33-39.

McKim, R. A. (1992), "Risk Management: Back to Basics," Cost Engineering, 34 #12 (December), pages 7-12.

Moore, Robert H. (1992), "Ethics and Risk Management," Risk Management, 39 #3 (March 1992):85-92.

Moss, V. (1992), "Aviation & Risk Management," Risk Management. 39 #7 (July), pages 10-18.

Petroski, H. (1994), Design Paradigms: Case Histories of Error & Judgment in Engineering. Cambridge U. Press, 1994.

Raftery, J. (1993), Risk Analysis in Project Management, Routledge, Chapman and Hall, London, U.K.

Schimrock, H. (1991), "Risk Management at ESA." ESA Bulletin. #67 (August), pages 95-98.

Sells, B. (1994), "What Asbestos Taught Me About Managing Risk," Harvard Business Review, 72 #2 (March/April), pages 76-90.

Shaw, T. E. (1990), "An Overview of Risk Management Techniques, Methods and Application." <u>AIAA Space Programs and Technology Conference Sept. 25-27</u>.

Smith, A. (1992), "The Risk Reduction Plan: A Positive Approach to Risk Management," <u>IEEE Colloquium on Risk Analysis Methods and Tools</u>.

Sprent, P. (1988), <u>Taking Risks: the Science of Uncertainty</u>, Penguin, New York, N.Y.

Stone, J. R. (1991), "Managing Risk in Civil Engineering by Machine Learning from Failures," <u>IEEE First International Symposium on Uncertainty Modeling and Analysis</u>, IEEE Computer Society Press, pages 255-259, Los Alamitos, CA.

Toft, B. (1994), <u>Learning From Disasters</u>, Butterworth-Heinemann, Woburn, Massachusetts.

Wideman, R. M. (1992), ed. <u>Project and Program Risk Management: A Guide to Managing Project Risks and Opportunities</u>, Project Management Institute, Drexel Hill, PA.

7

Project Scheduling and Budgeting Under Uncertainty

Accurate planning and effective scheduling are critical to the performance of any organization. Even the smallest project with limited resources has a huge number of combinations of resources and courses of action. The importance of a carefully considered plan and robust schedule cannot be overstated.

7.1 CASE STUDY: EXPANSION OF A FIBER OPTICS FIRM OFFICE SPACE

OpticalSoft is an established developer of fiber optical network software for PC's. Its recent growth has caused the need for rapid expansion of office space. Over the past several months, its programmers and systems staff have been spread out over three different buildings in a four-block downtown area. A major problem arises because people who need to be able to communicate with each other haven't been close enough for effective working relationships to develop. The company has decided to plan and build a new building very quickly.

OpticalSoft's management team has thus decided to use some of the ideas from "fast track" construction: begin building the rough parts of the building without knowing the final details about the interior. This requires a great deal of coordination, so information about finishing details are available when irrevocable construction activities are performed. The following activities were identified:

Figure 7.1 Project scheduling under uncertainty.

Site Selection and Purchase

If the new building is too far away from their present location, they'll lose some very talented people. Site selection can be started immediately. It is estimated that it will take 30 days to select and purchase the site. OpticalSoft has a good idea about the space they'll need and the land will be purchased with available cash. The activity will be called *SITE*.

Plans Preparation

OpticalSoft has already retained an architect who has assessed their needs and determined the requirements for the site. However, preparation of the plans can't begin until they have selected and purchased the site. Estimated time for plans preparation is 45 days. This activity will be called *PLANS*.

Revision of Plans

The experience of the architect indicated that 30 days should be allowed to revise plans after they have been prepared. This gives time to consider unforeseen governmental problems or problems that are detected when wider distribution of the plans is made. The name of this activity is *REVISE*.

Arrange Financing

The construction cannot be paid for with cash. Mortgage financing must be arranged. The mortgage will be secured with the land and building. This activity can begin as soon as the site purchase is completed. Estimated time to arrange financing is 60 days. The activity is called *FINANCE*.

Rough Construction

This phase of the construction involves site preparation, foundations, and structural elements. We'll begin on it as soon as the financing is complete and the plans are revised. Estimate time for rough construction is 90 days. The name of the activity is *ROUGH*.

Interior Design

Once rough construction is begun, the planning for detailed interior plans can begin. This includes the interior walls, plumbing, and electrical wiring. This shouldn't take too long, approximately 25 days. This activity will be called *INTERIOR DESIGN*.

Purchase Furnishings

As soon as the interior design is completed, the interior furnishings should be ordered. The furnishings include desks, tables, white boards, etc. This will take about 120 days because some of the items will need to be custom built. This activity will be called *FURNISH* or in some cases *FINISH*.

Interior Construction

This is the final stage of construction of the building. It can begin when two things have been finished: rough construction and completion of the computer network design. The best estimate of the time required for this activity is 70 days. The activity name is *INTERIOR CONSTRUCTION*.

PC Selection

While the site is being selected, the selection of the appropriate PC equipment can begin. This process will take a long time because everyone in the firm is a self-proclaimed expert. Approximately 120 days will probably be required. The name of this activity is *PC SELECT*.

PC Purchase

After the PC equipment has been selected, the actual equipment purchase can be made. OpticalSoft has enough clout with hardware vendors that good prices will be obtained. But since they think OpticalSoft is in a hurry, they won't budge on prices for a while. This activity will take about 90 days and will be called *PC PURCHASE*.

Network Design

The computing equipment must be interconnected through a network. At this stage, it is not clear what type of cabling will be used—CAT5 is a certainty, but whether it is coax or fiber has yet to be specified. Design of this stage cannot begin until the PC Equipment is selected; the interior construction cannot begin until the network is designed, because cabling must be in place before interior walls are finished. This design process will take 70 days. This activity is called *NETWORK*.

7.2 ESTABLISH A PROJECT SCHEDULING NETWORK

The activity diagram of the project is shown in Figure 7.2.

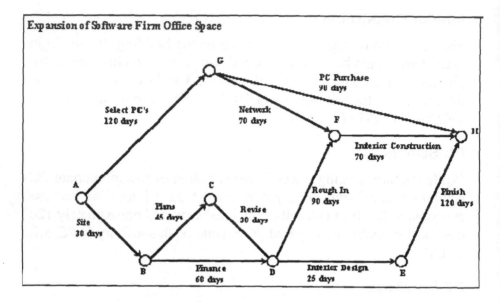

Figure 7.2 Establish a project scheduling network.

Definitions

Activity

An activity is a task that requires time to complete. Each activity is shown as an arc or arrow on a network. At each end of each activity is a node, which signifies an event. Associated with each activity is the time required to complete that activity. Each activity has a name associated with it, such as PC SELECT, NETWORK, etc.

Event

An event occurs when all of the activities pointing to that event have been completed; no activity which points from an event can begin until that event has occurred. Letters are used to name events. In the scenario above, we have events A through H; the project is complete when event H occurs.

Critical Path

The path through the network, such, that if the time for any activity in this path increases the project date will be delayed.

Crashing the Project

When the project critical path activities must be completed in a time frame less than the critical path, some activities on the critical path may be shortened by applying more resources to them, thus reducing the project time. This is called crashing the project.

Slack

Two types of slack exist--*free slack* and *total slack*. *Free slack* denotes the time than an activity can be delayed without delaying both the start of any succeeding activity and the end of the project. *Total slack* is the time that the completion of an activity can be delayed without delaying the end of the project.

Table 7.1 is developed based on the project scheduling network in Figure 7.2. The headings in Table 7.1 are described as follows:

L = length of an activity

Relationship of Events = the preceding and succeeding event for each activity

Early Start = the earliest time an activity can start without violating precedence relationships

Early Finish = the earliest time an activity can finish without violating any precedence relationships

Late Start = the latest time an activity can start without delaying completion of the project

Late Finish = the latest time an activity can finish without delaying the completion of the project

Free slack =the time that an activity can be delayed without delaying both the start of any succeeding activity and the end of the project

Total slack = the time that the completion of an activity can be delayed without delaying the end of the project.

Table 7.1 Prepare inputs for project scheduling analysis

ACTIVITY	Activity Length, L	Relationship	Early Start Time, ES=t	Early Finish EF=ES+t	Late Finish LF=T	Late Start LS=LF-L	Total Slack TS=LS-ES	Free Slack FS=tj-ti-L	Critical Activity
SITE	30	(A,B)	0	30	30	0	0	0	Critical
PC SELECT	120	(A,G)	0	120	125	5	5	0	
PLANS	45	(B,C)	30	75	75	30	0	0	Critical
FINANCE	60	(B,D)	30	90	105	45	15	15	
REVISE	30	(C,D)	75	105	105	75	0	0	Critical
INT DESIGN	25	(D,E)	105	130	145	120	15	0	
ROUGH	90	(D,F)	105	195	195	105	0	0	Critical
FINISH	120	(E,H)	130	250	265	145	15	15	
INT CONSTR	70	(F,H)	195	265	265	195	0	0	Critical
NETWORK	70	(G,F)	120	190	195	125	5	5	
PC PURCH	90	(G,H)	120	210	265	175	55	55	
Ending Event		H,X	265						

7.3 SPREADSHEET STRATEGIES FOR SOLVING CPM PROBLEMS USING EXCEL

There is a mechanism for determining the critical path for a project network without using the Excel Solver function. As shown in Figure 7.3, the process is an extension of the hand computations performed earlier, but has major implications for crashing a network and performing PERT computations.

1. The events are listed and named with a column for their scheduled time of occurrence. This is shown in cells A4:B12 on the sheet. No values should be filled in except for A, the starting event, which should be preset at 0. These are named using the INSERT NAME DEFINE command in Excel. These events were named A through H.
2. The activities were listed as shown, and columns created for:
    DURATION
    SCHEDULED START
    SCHEDULED FINISH
    ALLOWED BY SCHEDULE
    SLACK

	A	B	C	D	E	F	G
1							
2		**Spreadsheet Method for Determining Critical Paths**					
3		Earliest Time					
4	Event	of Occurrence		Cell Formulas for Column B			
5	A	0		Preset to 0			
6	B	30		=A+SITE			
7	C	75		=B+PLANS			
8	D	105		=MAX(C+REVISE,B+FINANCE)			
9	E	130		=D+INTERIOR DES			
10	F	195		=MAX(G+NETWORK,D+ROUGH)			
11	G	120		=A+PC SELECT			
12	H	265		=MAX(G+PC PURCH,F+INTERIOR CONSTR,E+FINISH)			
13							
14					Allowed		
15	Activity	Duration	Sched St	Sched Fin	By Sched	Slack	
16	SITE	30	0	30	30	0	
17	PLANS	45	30	75	45	0	
18	REVISE	30	75	105	30	0	
19	FINANCE	60	30	105	75	15	
20	ROUGH	90	105	195	90	0	
21	INTERIOR DES	25	105	130	25	0	
22	FINISH	120	130	265	135	15	
23	INTERIOR CONSTR	70	195	265	70	0	
24	PC SELECT	120	0	120	120	0	
25	PC PURCH	90	120	265	145	55	
26	NETWORK	70	120	195	75	5	

Figure 7.3 Spreadsheet method for determining critical path.

3. The activity times are entered into the DURATION column.
4. Compute the earliest time of occurrence for each event in B6:B12 (typically the first activity begins at 0; events with one preceding activity are the sum of the Early Start for that activity plus the Length of that activity; events with multiple preceding activities are the maximum of the Early Start + Length for each of the preceding activities). Note that the formulas used in column B are shown in column D.
5. The SCHEDULED START column is filled in by entering the name of the event that precedes that particular activity. For example PLANS would have +B as its contents, meaning that event B must precede the PLANS activity.
6. The SCHEDULED FINISH column is filled in by entering the name of the event that follows that particular activity. For

example PLANS would have +C as its contents, meaning that event C must follow the PLANS activity.

7. The ALLOWED BY SCHEDULE column is SCHED-ULED FINISH - SCHEDULED START.

8. SLACK (total slack) is simply ALLOWED BY SCHED-ULE - DURATION.

7.4 SOLVING CRITICAL PATH PROBLEMS USING EXCEL SOLVER

An efficient method for determining the critical path for a project network is to use the Excel Solver function to determine the critical path and the slack values.

The spreadsheet takes the problem posed earlier and hand cal-culated (with Excel assistance for arithmetic) and solves what is a linear programming solution to CPM. The following steps were used in the development of this spreadsheet.

1. The events are listed and named with a column for their scheduled time of occurrence. This is shown in cells B23:C31 on the sheet. No values should be filled in except for A_ the starting event. These are named using the IN-SERT NAME DEFINE command in Excel. These events were named A_ through H_.

2. The activities were listed as shown, and columns created for:
   DURATION
   SCHEDULED START
   SCHEDULED FINISH
   ALLOWED BY SCHEDULE
   SLACK

3. The activity times are entered into the DURATION column.

4. The SCHEDULED START column is filled in by entering the name of the event that precedes that particular activity. For example PLANS would have +B_ as its contents, mean-ing that event B must precede the PLANS activity.

5. The SCHEDULED FINISH column is filled in by entering the name of the event that follows that particular activity. For example PLANS would have +C_ as its contents, meaning that event C must follow the PLANS activity.

6. The ALLOWED BY SCHEDULE column is SCHEDULED FINISH - SCHEDULED START

7. SLACK (total slack) is simply ALLOWED BY SCHEDULE - DURATION

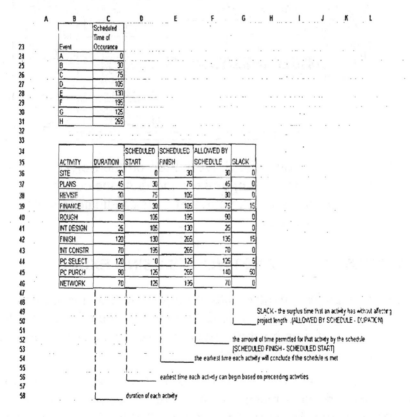

	A	B	C	D	E	F	G	H	I	J	K	L
			Scheduled Time of									
23		Event	Occurance									
24		A	0									
25		B	30									
26		C	75									
27		D	105									
28		E	130									
29		F	195									
30		G	125									
31		H	265									

	ACTIVITY	DURATION	SCHEDULED START	SCHEDULED FINISH	ALLOWED BY SCHEDULE	SLACK
36	SITE	30	0	30	30	0
37	PLANS	45	30	75	45	0
38	REVISE	30	75	105	30	0
39	FINANCE	60	30	105	75	15
40	ROUGH	90	105	195	90	0
41	INT DESIGN	25	105	130	25	0
42	FINISH	120	130	265	135	15
43	INT CONSTR	70	195	265	70	0
44	PC SELECT	120	0	125	125	5
45	PC PURCH	90	125	265	140	50
46	NETWORK	70	125	195	70	0

SLACK - the surplus time that an activity has without affecting project length (ALLOWED BY SCHEDULE - DURATION)

the amount of time permitted for that activity by the schedule [SCHEDULED FINISH - SCHEDULED START]

the earliest time each activity will conclude if the schedule is met

earliest time each activity can begin based on preceeding activities

duration of each activity

Figure 7.4 Using Excel Solver for determining critical path.

8. Solver selected from the TOOLS menu. The target cell is set for H (the concluding event) to be minimized. The cells that can change are C25:C31 (note that cell C24 is preset to

zero). The constraints are set to make sure that C25:C31 (the event times) remain positive and that G36:G46 (the slack times) also remain positive. This entry screen is shown below.

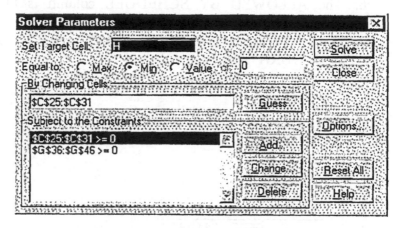

9. Finally, pressing the Solve button on the Solver Parameter screen activates the optimizer and finds the critical path and relevant slacks. Slack values of 0 are candidates for activities on the critical path.

7.5 CRASHING CPM NETWORKS USING EXCEL

Crashing a project involves attempting to reduce the completion time of a project from the time determined by the critical path to a lesser amount of time by applying additional resources to reduce the time of the most critical elements.

In general, one knows several things about each activity in the project network: how much it costs in resources (generally dollars) to reduce the time by a day, commonly called the *Daily Crash Cost*, and what the lower limit is that this rate per day will be valid, typically called the *Crash Limit*.

	A	B	C	D	E	F	G	H	I	J
1	CPM With Crash Optimizer Template for Excel									
2										
3		Sched. Time	Scratch							
4	Event	of Occurrence	Space							
5	A_	0								
6	B_	25	5							
7	C_	68	2							
8	D_	98	0							
9	E_	130	-6E-12							
10	F_	188	7.9E-13							
11	G_	118	-3E-12							
12	H_	250	1.2E-12							
13		0	8							
14		0	2							
15		0	-7E-12							
16		0	1.8E-15							
17		250	<--crash target							
18		107000	<--total crash cost							
19										
20			Scheduled	Scheduled		Allowed	Slack w/o	Days to	Crash	Slack With	Daily
21	Activity	Duration	Start	Finish	by Schedule	Crash	Crash	Limit	Crash	Crash Cost
22	SITE	30	0	25	25	-5	5	5	0	5000
23	PLANS	45	25	68	43	-2	2	2	0	500
24	REVISE	30	68	98	30	0	0	0	0	0
25	FINANCE	60	25	98	73	13	-6E-12	15	13	100
26	ROUGH	90	98	188	90	-1.4E-12	8E-13	10	-6E-13	10000
27	INTERIOR DES	25	98	130	32	7	-3E-12	5	7	50
28	FURNISH	120	130	250	120	5.66E-12	1E-12	10	6.9E-12	3000
29	INTERIOR CONS	70	188	250	62	-8	8	10	2.88E-13	10000
30	PC SELECT	120	0	118	118	-2	2	10	-4.4E-12	500
31	PC PURCH	90	118	250	132	42	7E-12	10	42	500
32	NETWORK	70	118	188	70	-1.4E-12	2E-15	10	-1.4E-12	500

Figure 7.5 Optimizing the solution for crashing a project.

The most efficient way to perform crashing in Excel is to use the Solver or optimizer function much as we did previously. In fact, the steps are similar. One should create a template as shown in Figure 7.5.

The steps to create this template follow:

1. The events are listed and named with a column for their scheduled time of occurrence. This is shown in cells B23:C31 on the sheet. No values should be filled in except for A_ the starting event. These are named using the INSERT NAME DEFINE command in Excel. These events were named A_ through H_. The column directly to the right of the event times should be labeled *Scratch Space*. In this

particular problem, we will need to have Solver modify the contents of B6:B12 and Solver will have to modify the contents of G22:G32. Unfortunately, Solver required that cells to be modified are continuous cells, in a row or in a block. Thus, the Scratch Space is used to indirectly modify the contents of G22:G32, with G22:G32 addressing the scratch space (e.g., cell G22's content is C6, its scratch location).

2. The activities were listed as shown, and columns created for:
   DURATION
   SCHEDULED START
   SCHEDULED FINISH
   ALLOWED BY SCHEDULE
   SLACK w/o CRASH
   DAYS TO CRASH
   CRASH LIMIT
   SLACK WITH CRASH
   DAILY CRASH COST

3. A location for the CRASH TARGET is found (B17). This is how many days in which we need to have the project completed.

4. A location for the TOTAL CRASH COST is found (B18). This is the location Solver reports the minimum expenditure needed to achieve the crash target. It uses the SUMPRODUCT function to multiply the pairwise products for the two columns DAYS TO CRASH and DAILY CRASH COST columns and sum those products.

5. The activity times are entered into the DURATION column.

6. The SCHEDULED START column is filled in by entering the name of the event that precedes that particular activity. For example PLANS would have +B_ as its contents, meaning that event B must precede the PLANS activity.

7. The SCHEDULED FINISH column is filled in by entering the name of the event that follows that particular activity. For example PLANS would have +C_ as its contents, meaning that event C must follow the PLANS activity.

8. The ALLOWED BY SCHEDULE column is SCHEDULED FINISH - SCHEDULED START.

9. SLACK W/O CRASH is simply ALLOWED BY SCHEDULE – DURATION.

10. DAYS TO CRASH is linked to the scratch area. For example cell G22 has as its contents +C6 indicating that its value is determined by Solver at cell C6. G23 has as its contents +C7 etc.

11. CRASH LIMIT is the predetermined minimum time that it takes to perform the particular activity under crash conditions. This is entered by the user.

12. SLACK WITH CRASH is the sum of the SLACK W/O CRASH and DAYS TO CRASH columns.

13. Finally, the DAILY CRASH COST is input by the user and represents the cost/day to perform at the crash level for this activity.

14. Solver is selected from the TOOLS menu. The Target Cell is set to minimize the total crash cost in cell B18. The cells that can change are B6:C16 and once again cell B5 is set to 0. This entry screen is shown below.

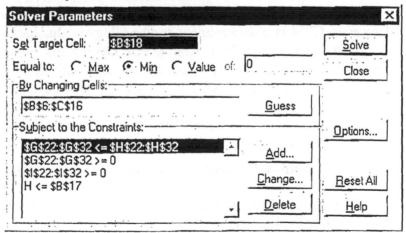

The constraints are set to make sure that the DAYS TO CRASH [COL G] are less than or equal to the CRASH LIMIT [COL H}; that DAYS TO CRASH remains positive [COL G>=0]; that the SLACKS WITH CRASH remains positive [COL I>=0]and, that H_ is restricted to be less than the target number of 250 days [H_<B17].

15. Finally, pressing the Solve button on the Solver Parameter screen activates the optimizer and finds the critical path, the relevant slacks, and the minimum cost to crash the project to 250 days. Slack values of 0 are candidates for activities on the critical path.

7.6 DEVELOPING AN APPROXIMATE PERT SOLUTION IN EXCEL

Program Evaluation and Review Technique (PERT) was developed by the consulting firm of Booz, Allen & Hamilton in conjunction with the United States Navy in 1958 as a tool for coordinating the activities of over 11,00 contractors involved with the Polaris missile program.

PERT is similar to CPM in that both are network oriented. However, CPM is a deterministic methodology in that each activity is considered of exact fixed length. PERT considers each activity stochastic in that variability is allowed in each activity. When one develops a PERT model, each activity is assigned three lengths:

a = an optimistic time (probability .9 of completion)

m = a typical time (modal)

b = a pessimistic time estimate (probability of .1 of completion)

The statistical model upon which PERT is based is known as the *Beta* distribution. The *Beta* distribution looks and behaves much like the normal distribution when m is exactly centered between a and b. When m is closer to a than to b, the *Beta* distribution becomes a positively skew unimodal distribution and when m is closer to b than to a, the *Beta* distribution becomes a negatively skew unimodal distribution. This difference allows a better fit of real-world data to activities than would the normal distribution. When activities that are Beta distributed are added, the resulting sum approaches the normal distribution which can be used to estimate completion times.

Several approached to PERT are possible. In the more complex modeling, each activity would be dealt with as a *Beta*-distributed variable, and the critical path(s) would be determined in that way. Notice the plural in the prior sentence on the word path. Because of the stochastic nature of each activity, the critical path can change as the length of activities change. When modeling a project network in this manner, the alternative CPM modeling strategy is used to determine the critical path.

The first PERT methodology considers *only* activities that have been previously determined to be on the critical path using a CPM model previously selected. In most cases, it is sufficient to provide reasonable probability estimates of project length. Advantages of this method include only having to develop three estimates for activities on the critical path and the inherent simplicity of the methodology.

1. Determine the critical path using CPM methodology. This could be done using either of the two CPM methodologies examined previously. Transfer only these activities to a different place on the spreadsheet or place them on another spreadsheet.

2 Estimate three times for each activity on the critical path.

   a = optimistic time (probability 0.9 of completion)

   m = typical time (modal)

   b = pessimistic time estimate (probability of 0.1 of completion)

   (These are the parameters for the requisite *Beta* distribution.)

	A	B	C	D	E	F	G	H	I
1	PERT ANALYSIS OF COMPLETION TIME FOR PROJECT								
2									
3									
4			PERT ESTIMATES				=(B8+4*C8+D8)/6		
5		Optimistic	Typical	Pessimistic	Expected				
6	Critical Activities	a	m	b	Duration	Variance	Std. Dev.		
7	SITE	15	30	40	29.17	17.36	4.17		
8	PLANS	40	45	55	45.83	6.25	2.50		
9	REVISE	25	30	40	30.83	6.25	2.50		
10	ROUGH	68	90	100	88.00	28.44	5.33		
11	INTERIOR CONST	60	70	80	70.00	11.11	3.33		
12									
13				SUM	263.83	69.42		=((+D10-B10)/6)^2	
14									
15	Enter Desired Completion Time		260						
16						=NORMDIST(C15,E13,SQRT(F13),TRUE)			
17	Estimation of Completion Probability			32.27%					

Figure 7.6 Developing an approximate PERT solution in Excel.

3. Determine the expected completion time, te, for each activity using the formula:

$$te = \frac{a+4m+b}{6}$$

4. Determine the completion time variance, v, for each activity using the formula:

$$v = \left[\frac{b-a}{6}\right]^2$$

5. Sum the expected completion times and the variances for all activities on the critical path. Denote the sum of the expected completion times as S, and the sum of the variances as V.

6. Compute the probability of completing on time by computing where D is the desired completion time. Use the normal distribution to determine the probability of completion.

$$Z = \frac{D-S}{\sqrt{V}}$$

The template shown in Figure 7.6 illustrates this methodology for the OpticalSoft sample case used.

7.7 DEVELOPING A COMPLETE PERT SOLUTION IN EXCEL

1. Create a model of the project network using the second (alternate) CPM methodology examined previously that uses the Excel logical *IF* and *MAX* functions in its construction.

2. Estimate three times for each activity on the critical path.

   a = optimistic time (probability 0.9 of completion)
   m = typical time (modal)
   b = pessimistic time estimate (probability of 0.1 of completion)

(These are the parameters for the requisite *Beta* distribution).

3. Determine the expected completion time, te, for each activity using the formula:

$$te = \frac{a+4m+b}{6}$$

4. Determine the completion time variance, v, for each activity using the formula:

$$v = \left[\frac{b-a}{6}\right]^2$$

5. Sum the expected completion times and the variances for all activities on the critical path. Denote the sum of the expected completion times as S, and the sum of the variances as V.

6. Compute the probability of completing on time by computing where D is the desired completion time. Use the normal distribution to determine the probability of completion.

$$ v = \left[\frac{b-a}{6} \right]^2 $$

	A	B	C	D	E	F	G	H	I	J	K	L	M	N	O	P	Q	R	S	T
1	PERT ANALYSIS OF COMPLETION TIME FOR PROJECT																			
2																				
3		Earliest Time	Variance																	
4	Event	of Occurrence	Name		Variance Cell Formulas for Column B							Cell Formulas for Col C								
5	A	0.00	VA	0								=VA+VSITE								
6	B	29.17	VB	17.36 =A+SITE								=VB+VPLANS								
7	C	75.00	VC	23.61 =B+PLANS								=F(C+REVISE>B+FINANCE,VC+VREVISE,VB+VFINANCE)								
8	D	135.83	VD	29.86 =MAX(C+REVISE,B+FINANCE)								=VD+VINTERIOR DES								
9	E	131.25	VE	31.22 =D+INTERIOR DES								=VA+VPC SELECT								
10	F	198.83	VF	58.31 =MAX(G+NETWORK,D+ROUGH)								=F(G+NETWORK>D+ROUGH,VG+VNETWORK,VD+VROUGH)								
11	G	119.17	VG	34.03 =A+PC SELECT								=F(G+PC PURCH>F+INTERIOR CONST)and(G+PC PURCH>E+FURNISH,								
12	H	258.83	VH	58.42 =MAX(G+PC PURCH,F+INTERIOR CONST,E+FURNISH)								VG+VPC PURC>,@IF+INTERIOR CONST>E+FURNSH,VF+VINTERIOR CONST,								
13												VE+VFURNISH))								
14		Enter Desired Completion Date in number must be between	247.17 and	280.50 ;--->	260.00															
15																				
16		Probability of Meeting Date	0.32																	
17				PERT ESTIMATES																
18		Expected			Optimistic Typical Pessimistic															
19		Activity	Duration	Variance Name	Variance Std. Dev	a	m	b												
20	SITE	29.17	VSITE	17.36	4.17	15	29	40												
21	PLANS	45.83	VPLANS	6.25	2.50	40	45	55												
22	REVISE	30.83	VREVISE	6.25	2.50	25	33	40												
23	FINANCE	30.83	VFINANCE	17.36	4.17	50	60	75												
24	ROUGH	88.00	VROUGH	28.44	5.33	65	90	100												
25	INTERIOR DES	25.50	VINTERIOR DES	1.36	1.17	23	25	30												
26	FURNISH	119.17	VFURNISH	34.03	5.83	100	120	135												
27	INTERIOR CONST	70.00	VINTERIOR CONST	11.11	3.33	60	70	80												
28	PC SELECT	119.17	VPC SELECT	34.03	5.83	100	120	135												
29	PC PURCH	31.67	VPC PURCH	25.00	5.00	80	80	110												
30	NETWORK	70.83	VNETWORK	17.36	4.17	60	70	85												

Figure 7.7 Developing a complete PERT solution in Excel.

The Excel template shown in Figure 7.7 illustrates this methodology for the OptiocalSoft sample case used.

In the CPM alternate method, the logical flow of the network was replicated with if statements to create all possible logical paths that the critical path might take, depending on the values. In this case, this logic must also be duplicated for the activity variances. The requisite statements are shown at the top of the sheet. The deterministic times in the CPM alternative method are replaced by the expected value computed from the three time estimates. To simplify formula creation, a great number of variables are named and used in these computations instead of the standard practice of using cell addresses like N56. When N56 is named COW, COW can be used in place of N56 in formulas to simplify them. This is accomplished by using the INSERT NAME tool in Excel.

7.8 PROBABILISTIC SOLUTIONS TO PROJECT SCHEDULING SIMULATION USING EXCEL

Monte-Carlo simulation provides an efficient probabilistic solution to project scheduling under uncertainty. The spreadsheet simulation process is summarized as follows:

1. Create a spreadsheet model of the project using activities and events.
2. Determine a random variate generator for each activity that is triangular, normal, uniform, or exponential use the generator to create random completion times for the activity.
3. Name as many variables as you can to name both the events and the activities.
4. Recreate the network logic next to the events using the =IF and MAX and arithmetic + operators to determine the project completion time.
5. Place the numbers 1-500 or 1-1000 down a column. This is the number of simulations you will be making.
6. In the cell to the right of, and just above, provide a link from the spreadsheet cell containing the project completion date as shown.
7. Create a data table using the DATA TABLE command in Excel and identify the range of the table as A30..B530 in this

example. Completing this command will produce a table that shows the different values that are shown in cell B30. When the spreadsheet asks you for the cell address of *column input cell*, give the address of any blank cell on the spreadsheet. When you have done this, the spreadsheet will simulate the completion of this project 500 times and place it in this newly created data table where it can be further analyzed. Analysis will yield information about how the completion time of the project. Excel provides the following help in carrying out this step.

	A	B	C	D	E	F	G	H	I	J	K	L	M	N
1	Probabilistic View of Critical Path Problem													
2														
3		Earliest Time												
4	Event	of Occurrence		Cell Formulas for Column B										
5	A	0												
6	B	36.7959		+A+SITE										
7	C	90.4726		+B+PLANS										
8	D	139.783		@MAX(C+REVISE,B+FINANCE)										
9	E	164.906		+D+INTERIOR DES										
10	F	238.514		@MAX(G+NETWORK,D+ROUGH)										
11	G	98.1653		+A+PC SELECT										
12	H	306.48		@MAX(G+PC PURCH,F+INTERIOR CONSTR,E+FURNISH)										
13														
14							Normal	Normal	Triangular	Triangular	Triangular	Triangular	Uniform	Uniform
15	Activity	Duration		Type of Distribution	Formula	Mean	STDEV	LOW	MODAL	HIGH	RAND	LLIM	UpLim
16	SITE	36.7959		Triangular (15,30,75)	36.7959			15	30	75	0.459424		
17	PLANS	53.6768		Normal (45,5)	53.6768	45	5						
18	REVISE	49.3105		Triangular (20,30,60)	49.3105			20	30	60	0.904779		
19	FINANCE	46.4237		Uniform (45,75)	46.4237							45	75
20	ROUGH	98.7309		Triangular (80,90,120)	98.7309			80	90	120	0.623021		
21	INTERIOR DES	25.123		Normal (25,3)	25.123	25	3						
22	FURNISH	114.85		Uniform (100,140)	114.85							100	140
23	INTERIOR CONS	67.9663		Triangular (60,70,100)	67.9663			60	70	100	0.159656		
24	PC SELECT	96.1653		Triangular (45,120,140)	96.1653			45	120	140	0.367423		
25	PC PURCH	93.5829		Triangular (60,90,100)	93.5829			60	90	100	0.897052		
26	NETWORK	34.3701		Normal (30,5)	34.3701	30	5						
27														
28	Simulation Results:			Statistical Summary of Simulation Results:										
29		306.48												
30	1	281.489												
31	2	321.643		max	360.16178		Enter Desired Completion Time		350					
32	3	285.297		min	232.60443		No Less Than		232.6044327					
33	4	278.645		avg	294.8487		No Greater Than		360.1617786					
34	5	289.85		stdev	19.813197									
35	6	290.117					Estimation of Completion Probability		99.00%					
36	7	329.954												
37	8	315.044												
38	9	294.133												
39	10	283.462		LL	LL	cum-freq	frequency	%f	cum %f					
40	11	303.769		220	240	1	1	0.2%	0.2%					
41	12	286.176		240	260	8	7	1.4%	1.4%					
42	13	288.766		260	280	129	121	24.2%	25.6%					
43	14	311.274		280	300	309	180	36.0%	61.6%					
44	15	295.271		300	320	448	139	27.8%	89.4%					
45	16	309.303		320	340	491	43	8.6%	98.0%					
46	17	298.417		340	360	499	8	1.6%	99.6%					
47	18	277.295		360	380	500	1	0.2%	99.8%					
48	19	313.377			TOTAL	2385	500							

Figure 7.8 Spreadsheet simulation using Excel.

8. Collect statistics from column B31..B530. Use the MAX, MIN, AVERAGE, and STDEV functions to find the maximum, minimum, and average completion times as well as the standard deviation of the completion times. You can build a frequency table using the Excel FREQUENCY function and approximate the probability of completion by using the PERCENTRANK function.

7.9 PROJECT BUDGETING

The goals of any project include bringing the project in on time and under or on budget. To accomplish these sometimes conflicting goals requires:

> a realistic initial project financial budget
> a feasible project time frame
> a budgetary plan linked to project activities
> an understanding of the relationship between completion time and project activities and their respective costs
> a methodology to track the variance between activity cost and budget

Each of these topics is an essential component of bringing a project in on time and within budget.

Creating an Initial Project Financial Budget

Project budgets are typically constructed in one of three ways:

> through top-down budget building

> through bottom-up budget building

> through a budget request process (combination of the above)

Top-Down Project Budget Building

Top-down budget development is based on the use of the collective judgments and experiences of top and middle managers and available past data when similar activities and projects were undertaken.

Typically, these cost estimates are then provided to middle management personnel who are expected to continue to breakdown these global estimates into specific work packages that comprise the subprojects. This process continues at respectively lower levels until the project budget is completed.

Disadvantages of Top-Down Budget Building

> subordinate managers often feel that they have insufficient budget allocations to achieve the objectives to which they must commit

> may cause unhealthy competition because this process is, in essence, a zero sum game -- one person's or area's gain is other's loss

Advantages of Top-Down Budget Building

> research shows that the aggregate budget is quite accurate, even though some individual activities may be subject to large error

> budgets are stable as a percent of total allocation and the statistical distribution of the budget is also stable leading to high predictability

> small, yet costly tasks, need not be identified early in this process--executive management experience has factored this into the overall estimate

Bottom-Up Budget Building

In bottom-up budget building elemental activities, tasks, their respective schedules, and their respective budgets are constructed through the work breakdown structure. The people doing the work are consulted regarding times and budgets for the tasks to ensure the best level of accuracy. Initially, estimates are made in terms of resources, such as man-hours and materials. These are later converted to dollar equivalents. Analytic tools such as learning curve analysis and work sampling are employed when appropriate to improve estimates. The individual components are then aggregated into a total project direct-cost project budget. Typically, the project manager then adds indirect costs (General & Administrative, or G&A), a project contingency reserve, and profit (if appropriate) to arrive at a total project budget.

Disadvantages of Bottom-Up Budget Building

individuals tend to overstate their resource needs because they suspect that higher management will probably cut all budgets by the same percentage

more persuasive managers sometimes get a disproportionate share of resources

a significant portion of budget building is in the hands of the junior personnel in the organization

sometimes critical activities are missed and left unbudgeted

Advantages of Bottom-Up Budget Building

leads to more accurate project budgeting because there is accurate costing at the activity level by those closest to the process

participation in the process leads to ownership and acceptance

Activity vs. Task-Oriented Budgets

The traditional organizational budget is activity oriented. Individual expenses are classified and assigned to basic budget lines such as phone, materials, personnel-clerical, utilities, direct labor, etc. These expense lines are gathered into more inclusive categories and are reported by organizational unit--for example, by section, department, or division. The budget is typically overlaid on the organizational chart.

With the advent of project organization, it became necessary to organize the budget in ways that conformed more closely to the actual pattern of fiscal responsibility. Under traditional budgeting methods, the budget for a project could be split up among many different organizational units. This diffused control so widely that it was frequently nonexistent. This problem led to program or task-oriented budgeting.

Program budgeting is the generic name given to a budgeting system that aggregates income and expenditures across programs (projects). In most cases, aggregation by program is in addition to, not instead of, aggregation by organizational unit. The project has its own budget. In the case of the pure project organization, the budgets of all projects are aggregated to the highest organizational level. When functional organization is used for projects, the functional department's budget is arranged in a manner standard to that organization, but income/expense for each project are shown. This is typically done on a spreadsheet with standard budget categories on the left side with totals, and the totals disaggregated into "regular operations" and charges to the various projects.

Building Feasible Project Time Frames

Just as the budget process produces project cost estimates either from top down, bottom up, or through a combination, project time frames can be determined. Top management may simply set an arbitrary completion date based on their collective wisdom. Or, the project completion date may be determined from the bottom up, after determining a preliminary project schedule. In many cases, project completion dates are externally determined--the new product must be introduced by September 12; the new student records system must be in place by August 1, etc.

When possible, once again, the combination method is probably the best method--a project completion date is established by senior management after a preliminary project schedule has been constructed. Thus, impossible or very difficult project completion dates may be avoided and large unnecessary expenditures may be avoided.

Linking Budgetary Planning to Project Activities

At some point, it is necessary to establish a linkage between project activities and the budget. Only in this way can a detailed budget plan be made. Further, it is the only way that the rate of depletion of the project budget can be monitored relative to activities completed. This phase should be done when the project schedule is determined.

Tracking the Variance (and Cumulative Variance) Between Activity Cost and Activity Budget

Reasonably, senior management want to know if projects are on time and under budget. This must be determined with some form of variance accounting. Most frequently, it is a cost report that lists activities, budget for the activity, the expenditure for the activity, the variance to budget, and whether the activity has been completed.

Column totals show the totals to date for these columns, and variance. Typically, the percent of the project completed is also shown, and the percent of the project budget consumed is also shown. Frequently, time estimates (+/-) are shown concurrent with that data along with a listing of major problem areas.

The Relationship Between Completion Time, Project Activities, and their Costs

There is a direct relationship between project length and project cost. The more time that must be cut from the duration of a project the more it will cost. Beyond a certain limit, no amount of additional resources will remove more time from a project.

Table 7.2 illustrates the places time may be found in shortening a project.

Table 7.2   Relationship Between Completion Time, Project
Activities, and their Costs

Typical Project Steps	Normal Project	Moderate Rush Project	Blitz Project
Project Approval Process	Full process	Somewhat abbreviated process	Only as necessary for major management decisions, purchasing and design engineering
Study of Alternatives	Reasonable	Quick study of major profitable items	Examine only those that do not affect schedule
Engineering Design	Begins near end of Approved Project Definition	Begins when Approved Project Definition is 50% to 75% complete	Concurrently with such Approved Project definition as is done
Issue Engineering to Field	Allow adequate time for field to plan and purchase field items	Little no lead time between issued and field erection	No lead time between issue and field erection
Purchasing	Begins in later stages of Approved Project Definition	Done concurrently with Approved Project Definition. Rush purchase of all long delivery items.	Done concurrently with Approved Project Definition. Rush purchase of all long delivery items. Rush buy anything that will do the job. Over-order.
Premium Payments	Negligible	Some to break specific Bottlenecks	As necessary to forestall any possible delays
Field Crew Strength	Minimal practical or optimal cost	Large crew with some spot overtime.	Large crew; overtime and/or extra shifts
Probable Cost Difference Design & Development Engineering & Construction	Base level	5% to 10% more 3% to 5% more	15% and up more 10% and up more
Probable Time	Base level	up to 10% less	Up to 50% less

7.10 SUMMARY

A project contains interrelated activities. Each of interrelated activities has a definite starting and ending point, and results in a unique product or service. Project scheduling is goal oriented. Project planning and scheduling are common to many different engineering domains. But whether the project is as large as the Boston Harbor Tunnel or something as seemingly simple as the redesign of the packaging for a tape dispenser, both planning and scheduling are profoundly important. Even on a small project, the number of possible courses of action and the number of ways to allocate resources quickly become overwhelming.

Many methods are used for project planning. Perhaps the most common planning tool is Project Evaluation and Review Technique (PERT). PERT specifies technological or other types of precedence constraints among tasks in a project. Work break down structure (WBS) organizes tasks into functional groups. Other methods such as "four fields" focus on team members or project components rather than simply precedence relationships. Most project planning tools include more than one view of the project; a typical project planner will present a PERT chart to show precedence relationships, a WBS chart to show hierarchical or technological relationships, and some kind of temporal view, typically a Gantt chart, to illustrate estimated and actual start and completion times. In addition, many different resource usage reports are typically generated. For descriptions of these methods, see (NASA, 1962) (Kelley, 1959) (Willis, 1985) (Scasso, 1991) (O'Sullivan, 1991).

Scheduling methods typically start with a project plan then assign start times given temporal, precedence, or resource constraints. For example, one commonly used scheduling technique, the Critical Path Method (CPM), starts with a PERT chart then schedules tasks based on the precedence relationships and estimates of task duration in the PERT representation. It is important to make a distinction between planning and scheduling. Using traditional tools, the PERT chart represents the project plan whereas the Gantt chart represents the project schedule. Often scheduling and planning are not done at the same time; scheduling is much easier once a plan has been established, and thus typically follows planning. As a result, planning and scheduling are often done in an iterative manner (or in some cases, sequentially with no iteration). But optimal planning can require scheduling information. For example, if a key resource will not be available during a certain period, then one may want to change the plan to take advantage of a different resource, possibly using a different process. Although scheduling usually cannot be done until a plan is established, in cases with highly constrained resources, the plan cannot be established until the schedule is (at least partially) established. In general, the more constrained the resources, the more tightly coupled planning and scheduling will be.

REFERENCES

Boctor, F. F. (1993), "Heuristics for scheduling projects with resource restrictions and several resource-duration modes," International Journal of Production Research, 31(11), pages 2547.

Boctor, F. F. (1994), "An adaptation of the simulated annealing algorithm for solving resource-constrained project scheduling problems," Document de Travail 94-48. Groupe de Recherche en Gestion de la Logistique, Université Laval, Québec, Canada.

Crandall, K. C. (1973), "Project planning with precedence lead/lag factors," Project Management Quarterly, 4(3), pages 18-27.

Davis, E. W. (1973), "Project Scheduling Under Resource Constraints Historical Review and Categorization of Procedures," AIIE Transactions (November), pages 297-313.

Davis, E. W. and J. H. Patterson (1975), "A Comparison of Heuristic and Optimum Solutions in Resource-Constrained Project Scheduling," Management Science 21(8), pages 944-955.

Davis, K. R. (1992). "Resource Constrained Project Scheduling With Multiple Objectives: A Decision Support Approach." Computers and Operations Research, 19(7), page 657.

Icmeli, O. (1993), "Project Scheduling Problems: A Survey," International Journal of Operations and Production Management, 13(11), page 80.

Jeffcoat, D. E. (1993), "Simulated annealing for resource-constrained scheduling," European Journal of Operational Research, 70(1), page 43.

Johnson, R. V. (1992), "Resource Constrained Scheduling Capabilities of Commercial Project Management Software," Project Management Journal, XXII(4), page 39.

Kelley, J. E. Jr. and M. R. Walker (1959), "Critical Path Planning and Scheduling," Proceedings of the Eastern Joint Computer Conference, pages 160-173.

Li, R. K.-Y. and R. J. Willis (1993), "Resource Constrained Scheduling Within Fixed Project Durations," OR: The Journal of the Operational Research Society, 44(1), page 71.

NASA (1962), PERT/Cost System Design, DOD and NASA Guide, Washington DC, Office of the Secretary of Defense, National Aeronautics and Space Administration.

O'Sullivan, D. (1991), "Project management in manufacturing using IDEF0," International Journal of Project Management, 9(3), page 162.

Patterson, J. H. (1984), "A Comparison of Exact Approaches for Solving the Multiple Constrained Resource, Project Scheduling Problem," Management Science 30(7), page 854.

Sampson, S. E., and E. N. Weiss (1993), "Local Search Techniques for the Genralized Resource Constrained Project Scheduling Problem," Naval Research Logistics, 40(5), page 665.

Scasso, R. d. H. and G. S. Larenas (1991), "Project-breakdown structure: the tool for representing the project system in project management," International Journal of Project Management, 9(3), page 157.

Software Digest (1993), "Project Management Software Reviews.," Software Digest, Ratings Report. 10.

Talbot, F. B. (1982), "Resource-Constrained Project Scheduling with Time-Resource Tradeoffs: the Nonpreemtive Case," Management Science, 28(10), pages 1197-1211.

Tavares, L. V. and J. Weglarz (1990), "Project Management and Scheduling: A Permanent Challenge for OR," European Journal of Operational Research, 49(1-2).

Weglarz, J. (1979), "Project scheduling with discrete and continuous resources," IEEE Transactions on Systems, Man, and Cybernetics, 9(10), pages 644-650.

Weglarz, J. (1981), "Project scheduling with continuously-divisible, doubly constrained resources," Management Science, 27(9), pages 1040-1053.

Willis, R. J. (1985), "Critical path analysis and resource constrained project scheduling - theory and practice," European Journal of Operational Research, 21, pages 149-155.

8

Process Control – Decisions Based on Charts and Index

The concept of quality has been with us since the beginning of time. As early as the creation of the world described in the Bible in Genesis, God pronounced his creation "good"-- e.g., acceptable quality. Artisans' and craftsmen's skills and the quality of their work are described throughout history. Typically the quality intrinsic to their products was described by some attribute of the products such as strength, beauty or finish. However, it was not until the advent of the mass production of products that the reproducibility of the size or shape of a product became a quality issue.

8.1 PROCESSES AND PROCESS VARIABILITY

Quality, particularly the dimensions of component parts, became a very serious issue because no longer were the parts built hand and individually fitted until the product worked. Now, the mass-produced part had to function properly in every product built. Quality was obtained by inspecting each part and passing only those that met specifications. This was true until 1931 when Walter Shewhart, a statistician at the Hawthorne plant at Western Electric, published his book *Economic Control of Quality of Manufactured Product* (Van Nostrand, 1931). This book is the foundation of modern statistical process control (SPC) and provides the basis for the philosophy of total quality management or continuous process improvement for improving processes. With statistical process control, the process is monitored through sampling. Considering the results of the sample, adjustments are made to the process before the process is able to produce defective parts.

The concept of process variability forms the heart of statistical process control. For example, if a basketball player shot free throws in practice, and the player shot 100 free throws every day, the player would not get exactly the same number of baskets each day. Some days the player would get 84 of 100, some days 67 of 100, some days 77 of 100, and so on. All processes have this kind of variation or variability.

The capability of a process is defined as the inherent variability of a process in the absence of any undesirable special causes; the smallest variability of which the process is capable with variability due solely to common causes.

Typically, processes follow the normal probability distribution. When this is true, a high percentage of the process measurements fall between $\pm 3\sigma$ of the process mean or center. That is, approximately 0.27% of the measurements would naturally fall outside the $\pm 3\sigma$ limits and the balance of them (approximately 99.73%) would be within the $\pm 3\sigma$ limits.

Since the process limits extend from -3σ to $+3\sigma$, the total spread amounts to about total variation. If we compare process spread with specification spread, we typically have one of three situations:

Case I   A Highly Capable Process

The Process Spread is Well Within the Specification

Spread

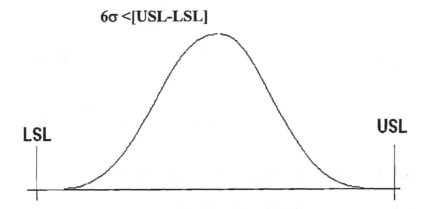

$$6\sigma < [USL-LSL]$$

LSL   USL

Figure 8.1 The Process Spread is Well Within the Specification

As illustrated in Figure 8.1, we have an attractive situation for several reasons when processes are capable. We could tighten our specification limits and claim our product is more uniform or consistent than our competitors. We can rightfully claim that the customer should experience less difficulty, less rework, more reliability, etc. This should translate into higher profits.

Case II   The Process Spread Just About Matches

$$6\sigma = [USL-LSL]$$

As illustrated in Figure 8.2, when a process spread is just about equal to the specification spread, the process is capable of meeting specifications, but barely so. This suggests that if the process mean moves to the right or to the left just a little bit, a significant amount of the output will exceed one of the specification limits. The process must be watched closely to detect shifts from the mean. Control charts are excellent tools to do this.

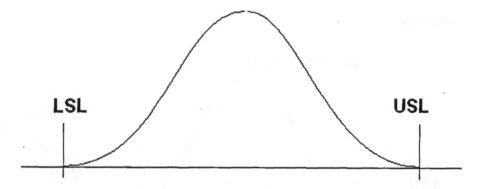

Figure 8.2 The Process Spread Just About Matches

Case III   A "Not Capable" Process

The Process Spread Is Within 6σ

$$6\sigma > [USL\text{-}LSL]$$

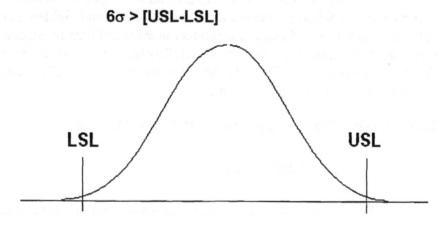

Figure 8.3 The Process Spread Is Within 6σ.

As illustrated in Figure 8.3, when the process spread is greater than the specification spread, a process is not capable of meeting specifications regardless of where the process mean or center is located. This is indeed a sorry situation. Frequently this happens, and the people responsible are not even aware of it. Over adjustment of the process is one consequence, resulting in even greater variability. Alternatives include:

- Changing the process to a more reliable technology or studying the process carefully in an attempt to reduce process variability.
- Live with the current process and sort 100% of the output.
- Re-center the process to minimize the total losses outside the spec limits
- Shut down the process and get out of that business.

8.2 STATISTICAL PROCESS CONTROL

Shewhart's discovery of statistical process control or SPC, is a methodology for charting the process and quickly determining when a process is "out of control" (e.g., a special cause variation is present because something unusual is occurring in the process). The process is then investigated to determine the root cause of the "out of control" condition. When the root cause of the problem is determined, a strategy is identified to correct it. The investigation and subsequent correction strategy is frequently a team process and one or more of the TQM process improvement tools are used to identify the root cause. Hence, the emphasis on teamwork and training in process improvement methodology.

It is management's responsibility to reduce common cause or system variation as well as special cause variation. This is done through process improvement techniques, investing in new technology, or reengineering the process to have fewer steps and therefore less variation. Management wants as little total variation in a process as possible--both common cause and special cause variation. Reduced variation makes the process more predictable with process

output closer to the desired or nominal value. The desire for abso-
lutely minimal variation mandates working toward the goal of reduced process variation.

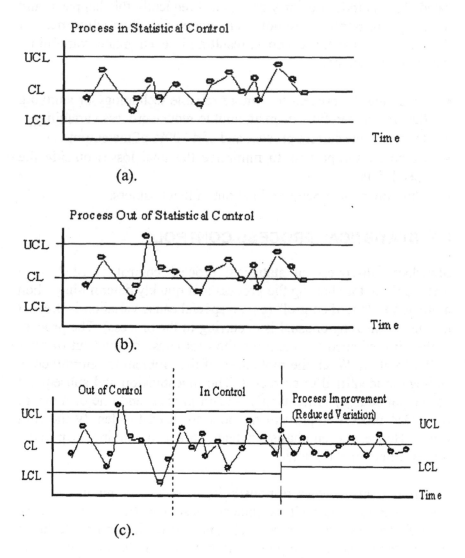

Figure 8.4 Statistical process control: three types of processes.

The process shown in Figure 8.4(a) is in apparent statistical control. Notice that all points lie within the upper control limits (UCL) and the lower control limits (LCL). This process exhibits only common cause variation.

The process shown in Figure 8.4(b) is out of statistical control. Notice that a single point can be found outside the control limits (above them). This means that a source of special cause variation is present. The likelihood of this happening by chance is only about 1 in 1,000. This small probability means that when a point is found outside the control limits that it is very likely that a source of special cause variation is present and should be isolated and dealt with. Having a point outside the control limits is the most easily detectable out of control condition.

The graphic shown in Figure 8.4(c) illustrates the typical cycle in SPC. First, the process is highly variable and out of statistical control. Second, as special causes of variation are found, the process comes into statistical control. Finally, through process improvement, variation is reduced. This is seen from the narrowing of the control limits. Eliminating special cause variation keeps the process in control; process improvement reduces the process variation and moves the control limits in toward the center line of the process.

8.3 TYPES OF OUT-OF-CONTROL CONDITIONS

Several types of conditions exist that indicate that a process is out of control. The first of these we have seen already—having one or more points outside the $\pm 3\sigma$ limits as shown below:

Extreme Point Condition

This process is out of control because a point is either above the UCL or below the UCL.

Process Out of Statistical Control [point(s) outside control limits]

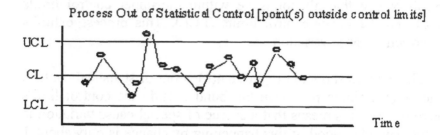

Figure 8.5 Process Out of Statistical Control.

This is the most frequent and obvious out of control condition and is true for all control charts (see Figure 8.5).

Control Chart Zones

As shown in Figure 8.6, Control charts can be broken into three zones, a, b, and c on each side of the process center line.

A series of rules exist that are used to detect conditions in which the process is behaving abnormally to the extent that an out of control condition is declared.

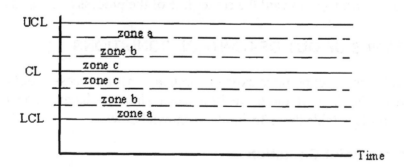

Figure 8.6 Control Chart Zones.

Two of Three Consecutive Points in Zone A or Outside Zone A

The probability of having two out of three consecutive points either in or beyond zone A is an extremely unlikely occurrence when the process mean follows the normal distribution. Thus, this criteria applies only to charts for examining the process mean.

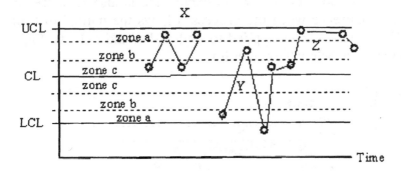

Figure 8.7 Two of Three Consecutive Points in Zone A or Outside Zone A.

X, Y, and Z in Figure 8.7 are all examples of this phenomena.

Four of Five Consecutive Points in Zone B or Beyond

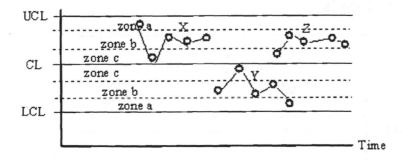

Figure 8.8 Four of Five Consecutive Points in Zone B or Beyond Runs Above or Below the Center Line.

The probability of having four out of five consecutive points either in or beyond zone B is also an extremely unlikely occurrence when the process mean follows the normal distribution. Again this criteria should only be applied to an chart when analyzing a process mean. X, Y, and Z in Figure 8.8 are all examples of this phenomena.

The probability of having long runs (8 or more consecutive points) either above or below the center line is also an extremely unlikely occurrence when the process follows the normal distribution. This criteria can be applied to both X-bar and r charts. Figure 8.9 shows a run below the center line.

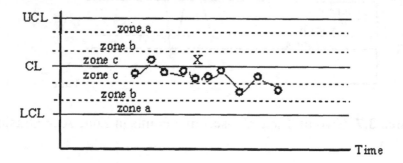

Figure 8.9 A Run (8 or More Consecutive Points) Below the Center Line.

Linear Trends

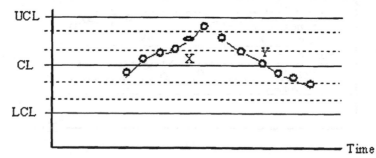

Figure 8.10 The Linear Trends.

The probability of 6 or more consecutive points showing a continuous increase or decrease is also an extremely unlikely occurrence when the process follows the normal distribution. This criteria can be applied to both X-bar and r charts. As shown in Figure 8.10, X and Y are both examples of trends. Note that the zones play no part in the interpretation of this out of control condition.

Oscillatory Trend

The probability of having 14 or more consecutive points oscillating back and forth is also an extremely unlikely occurrence when the process follows the normal distribution. It also signals an out of control condition. This criteria can be applied to both X-bar and r charts.

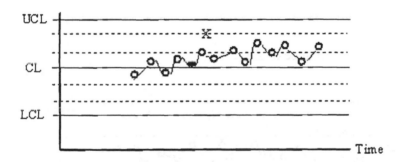

Figure 8.11 Oscillatory Trend

X in Figure 11 is an example of this out of control condition. Note that the zones play no part in the interpretation of this out of control condition.

Avoidance of Zone C

The probability of having 8 or more consecutive points occurring on either side of the center line and do not enter Zone C is also an extremely unlikely occurrence when the process follows the normal distribution and signals an out of control condition. This criteria can be applied to X-bar charts only. This phenomena occurs when more than one process is being charted on the same chart (probably by accident—e.g., samples from two machines mixed and put on a single chart), the use of improper sampling techniques, or perhaps the process is over controlled or the data is being falsified by someone in the system. X in Figure 12 is an example of this out of control condition.

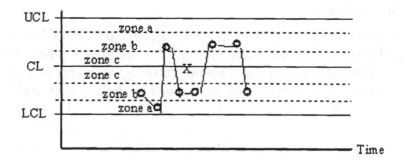

Figure 8.12 Avoidance of Zone C.

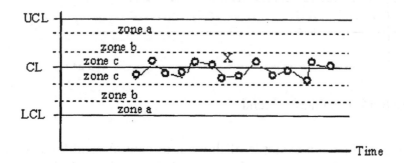

Figure 8.13 Run in Zone C.

Run in Zone C

The probability of having 15 or more consecutive points occurring the Zone C is also an extremely unlikely occurrence when the process follows the normal distribution and signals an out of control condition. This criteria can be applied to charts only. This condition can arise from improper sampling, falsification of data, or a decrease in process variability that has not been accounted for when calculating control chart limits, UCL and LCL. X in Figure 13 is an example of this out of control condition.

8.4 SAMPLING AND CONTROL CHARTS

Several issues are important when selecting sample for control chart purposes. They include:

- sample (subgroup) size considerations
- sampling frequency
- collecting samples

A major goal when selecting a sample from a process is to select the sample (subgroup) in such a way that the variation within the subgroup is attributable only to the random variation inherent in the process or common cause variation. The idea is that the sample should be chosen in such a manner that the chances are maximized to have each element in the sample be alike and subject only to common cause variation. The *spacing* of the samples (subgroups) is arranged so that *if* special cause variation is present, the control chart can identify its presence.

Considerations in Determining Subgroup (Sample) Size

- Subgroups should be small enough to be economically and practically feasible. The time and effort it takes to collect and measure samples weighs heavily here.
- Subgroups should be large enough to allow the central limit theorem to cause the distribution of sample means to be normally distributed. In many cases the process measurements are normal, or close to normal. In a few cases they may not be normal. We know from the central limit theorem that the larger the sample size, the more likely it is that the distribution of sample means will follow the normal distribution. From a practical perspective, this is true for most subgroup sizes of four or more.

- Larger subgroups also are needed to provide good sensitivity in detecting out of control conditions. The larger the subgroup size, the more likely it is that a shift in a process mean would be detected.

- As mentioned previously, subgroups should be selected so that they are subject ONLY to common cause variation. If subgroups are allowed to get very large, it is possible that special cause variation can be mixed with common cause. This effect will reduce the sensitivity of the control chart in detecting shifts in the process characteristic of interest.

When all of the above considerations is taken into account, typically a subgroup size of between 4 and 6 is selected. Five is the most commonly used subgroup size (this is due to the fact that since 5 is half of 10, many computations with 5 can be done mentally - with calculators and computers, it is probably no longer an important consideration).

Considerations in Collecting Samples

Typically, we want measurement within a subgroup to be taken as close to the same time as possible to reduce the change that special cause variation is present within the subgroup. Thus it is common that consecutive samples from a process are taken. A period of time elapses, and another subgroup sample is collected consecutively.

The spacing between the subgroups shouldn't be exactly uniform. It is not a good idea to take samples EXACTLY every hour or at EXACTLY the same time each day. A certain amount of randomness in the interval between samples is good because it tends to minimize the effect of shift changes, tool wear, tool changes, etc. If the rule is to take samples hourly, a better plan might be to take them hourly, but vary the time randomly within ± 10 minutes of the hour interval.

The Frequency of Sample Collecting

The bottom line is that samples must be collected frequently enough to be useful in identifying and solving problems. In many cases in industry, samples are collected too infrequently. The following should be considered:

- **Process Stability** - if a process has not been analyzed using control charts before and exhibits erratic behavior, sampling should be more frequent to increase the opportunities for process improvement. In this case, frequently ALL parts are sampled, measured, and grouped serially into groups of 5, for example, and then charted. The frequency between these samples of 5 is reduced as the process becomes more stable.
- **Frequency of Process Events** - If a process has may things happening in it, material changes, tool changes, process adjustments, etc., sampling should take place after these potential special causes so they can be detected. When many special events are taking place in a process, each shift, taking two samples (subgroups) per shift will be of little benefit.
- **Sampling Cost** - Two considerations can occur. The time involved in taking the sample is one factor and if the quality characteristic can be observed only through destructive testing, the loss or output can be a significant cost. These costs must be weighed when determining the frequency of sampling. A much more usual condition is that the sampling cost is deemed too high and the frequency of sampling is reduced to the level that the charts derived have no value. In this case, a great deal is spent on sampling with no value derived from the charts and ALL of the expenditures are wasted. Thus, if the process is to be sampled, the samples should be taken frequently enough that the resulting charts are of value. Otherwise, charting these processes should simply be abandoned.

The Problems of Stratification and Mixing

Stratification occurs when the output of several parallel (and assumed identical) processes into a single sample for charting the combined process. Typically a single sample is taken from each machine and included in the subgroup. If a problem develops in the process for one of the machines, it is very difficult to detect because the sample from the "problem" machine is grouped with other samples from "normal" machines. What is plotted as common cause variation is really common cause variation plus the slight differences between the process means of the individual machines. Typically, stratification is detected when large numbers of points lie in Zone C of an chart. It looks like the process is in super control when, in fact, the control limits are just calculated too wide. The solution to stratification, obviously, is to chart each machine separately. Control charts are applicable to one and only one process at a time.

Mixing is similar to stratification, except the output of several parallel machines is mixed and the periodic sample is drawn from the mixture. Similar to stratification, mixing will mask problems in individual machines and will make isolation of the problem difficult. Mixing tends to produce an appearance on the control chart where points tend to lie closer to the control limits than they really should be. The more dissimilar the machines, the more pronounced this phenomena will be. Frequently mixtures come from processes such as multispindle screw machines and multicavity molds. The solution to mixing, obviously, is to chart each machine or mold separately. Control charts are applicable to one and only one process at a time.

8.5 STEPS IN DETERMINING PROCESS CAPABILITY

Determine if Specifications Are Currently Being Met

- Collect at least 100 random samples from the process.
- Calculate the sample mean and standard deviation to estimate the true mean and standard deviation of the process.
- Create a frequency distribution for the data and determine if it is close to being normally distributed. If it is continue; if not, get the help of a statistician to transform the data or to find an alternative model.

- Plot the USL and LSL on the frequency distribution.
- If part of the histogram is outside the specification limits, consider adjusting the mean to center the process.
- If the histogram indicates that the process spread is greater than the specification spread, the process might not be capable.

Determine the Inherent Variability Using an r-Chart

1. Get at least 40 rational subgroups of sample size, preferably at least 4 or 5.
2. Calculate the ranges for each subgroup, the average range, rbar, and the control limits for an r-chart. Plot the data.
3. Discard any ranges outside the UCL ONLY if the undesirable special cause is identifiable and can be removed from the process; otherwise include the offending range(s).
4. Recalculate the average range, rbar, and the control limits.
5. Repeat the last two steps until all ranges are in statistical control.
6. Estimate the process standard deviation, σ.
7. Using the midpoint of the specifications as the process mean, assuming normality, draw a normal curve, and estimate the percentage meeting specifications. This step assumes that the mean of the process can be adjusted or recentered.
8. If the normal curve shows that the specifications can be met, determine what the specifications are not being met. Create an X-bar chart from the same subgroup data and hunt for clues. It may be as simple as recentering the process. Perhaps special causes are present that can be removed.
9. If the specifications cannot be met, consider changing the process by improvement, living with it and sorting 100% of the output, centering the mean, or dropping the product totally.
10. Set up a system of X bar-r charts to create future process improvements and process control.

8.6 CAPABILITY INDEXES

Capability indexes are simplified measures to quickly describe the relationship between the variability of a process and the spread of the specification limits. Like many simplified measures, such as the grades A, B, C, D, and F in school, capability indexes do not completely describe what is happening with a process. They are useful when the assumptions for using them are met to compare the capabilities of processes.

The Capability Index - C_p

The equation for the simplest capability index, C_p, is the ratio of the specification spread to the process spread, the latter represented by six standard deviations or 6σ.

$$C_p = \frac{USL - LCL}{6\sigma}$$

It assumes that the normal distribution is the correct model for the process. It can be highly inaccurate and lead to misleading conclusions about the process when the process data does not follow the normal distribution.

Occasionally the inverse of the capability index C_p, the capability ratio CR is used to describe the percentage of the specification spread that is occupied or used by the process spread.

$$CR = CapabilityRatio$$

$$= \frac{1}{Cp} \times 100\%$$

$$= \frac{6\sigma}{USL - LSL} \times 100\%$$

C_p can be translated directly to the percentage or proportion of nonconforming product outside specifications.

- When C_p =1.00, approximately 0.27% of the parts are outside the specification limits (assuming the process is centered on the midpoint between the specification limits) because the specification limits closely match the process UCL and LCL. We say this is about 2700 parts per million (ppm) nonconforming.

- When C_p =1.33, approximately 0.0064% of the parts are outside the specification limits (assuming the process is centered on the midpoint between the specification limits). We say this is about 64 parts per million (ppm) nonconforming. In this case, we would be looking at normal curve areas beyond $1.33x(\pm 3\sigma) = \pm 4\sigma$ from the center.

- When C_p =1.67, approximately 0.000057% of the parts are outside the specification limits (assuming the process is centered on the midpoint between the specification limits). We say this is about 0.6 parts per million (ppm) nonconforming. In this case, we would be looking at normal curve areas beyond 1.67x= from the center of the normal distribution. Remember that the capability index ignores the mean or target of the process. If the process mean lined up exactly with one of the specification limits, half the output would be nonconforming regardless of what the value of was. Thus, is a measure of potential to meet specification but says little about current *performance* in doing so.

The Capability Index - C_{pk}

The major weakness in C_p was the fact that few, if any processes remain centered on the process mean. Thus, to get a better measure of the current performance of a process, one must consider where the process mean is located relative to the specification limits. The index C_{pk} was created to do exactly this. With C_{pk} , the location of the process center compared to the USL and LSL is included in the computations and a worst case scenario is computed in which C_{pk} is computed for the closest specification limit to the process mean.

$$C_{pk} = \min\left\{\frac{USL - \mu}{3\sigma} \; and \; \frac{\mu - LS :}{3\sigma}\right\}$$

As shown in Figure 14, we have the following situation. The process standard deviation is $\varsigma=0.8$ with a USL=24, LSL=18, and the process mean �=22.

$$C_{pk} \quad \min \left[\frac{24 \quad 22}{3\upsilon 8} \ and \ \frac{22 \quad 18}{3\upsilon 8} \right]$$

$$= \min\{0.83 \ and \ 1.67\}$$

$$= 0.83$$

If this process' mean was exactly centered between the specification limits,

$$C_p \quad C_{pk} \quad 1.25$$

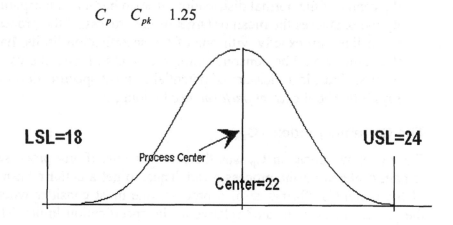

LSL=18   USL=24

Process Center

Center=22

Figure 8.14 The Capability Index - C_{pk}

The Capability Index - C_{pm}

C_{pm} is called the Taguchi capability index after the Japanese quality guru, Genichi Taguchi whose work on the Taguchi Loss Function stressed the economic loss incurred as processes departed from target values. This index was developed in the late 1980's and takes into account the proximity of the process mean to a designated target, T.

$$C_{pm} = \frac{USL - LSL}{6\sqrt{\sigma^2 + (\mu - T)^2}}$$

When the process mean is centered between the specification limits and the process mean is on the target, T, $C_p = C_{pk} = C_{pm}$.

When a process mean departs from the target value T, there is a substantive affect on the capability index. In the C_{pk} example above, if the target value were T=21, would be calculated as:

$$C_{pm} = \frac{24 - 18}{6\sqrt{8^2 + (22 - 21)^2}}$$

In this case, the Taguchi capability index is somewhat more liberal than C_{pk}.

8.7 CONTINUOUS QUALITY IMPROVEMENT VS. BUSINESS PROCESS RE-ENGINEERING

It is worthwhile to briefly describe the differences between the philosophies and practices of Continuous Quality Improvement (CQI) or Total Quality Management (TQM) and Business Process Reengineering (BPR). For purposes of this discussion, we will use the terms CQI and TQM interchangeably.

What is CQI and/or TQM?

The philosophy of TQM or CQI is best credited to two people, Walter Shewhart who was a statistician at the Western Electric Hawthorne plant in the 1930's and his protégé, W. Edwards Deming. It was Shewhart who developed the concept of statistical process control (SPC) while attempting to improve the quality of telephones made at the Hawthorne plant. As we have seen earlier, the basic concept is to use a system of charting to separate ordinary process variation (common cause) from variation due to unusual circumstances in the process (special cause variation). When the system of charting indicates that a special cause has occurred, the process is analyzed to determine the cause of this variation and it is remedied. Then the process is restarted.

When a process has been refined and has small common cause variability, SPC can detect process problems and cause them to be repaired BEFORE the process produces any bad parts. The "bad" part, or series of "bad" parts that triggered the out of control condition are still within the specification limits and are therefore good enough to be used.

Table 8.1 A Table of Differences Contrasting CQI/TQM and BPR

Variable	CQI/TQM	BPR
Organization	Common goals across functions	Process based
Focus	Quality; attitude toward customers	Processes; minimize non-value added steps
Improvement Scale	Continual; incremental	Radical change
Customer Focus	Internal and external satisfaction	Outcomes driven
Process Focus	Simplify; improve; measure to achieve control	Optimal streamlined system of processes with few steps not adding value
Techniques	Process maps; benchmarking; self-assessment; SPC; PDCA; quality tools	Process maps; benchmarking; information technology; breakthrough thinking; re-engineering tools

For SPC to work effectively, the common cause variation must be small relative to the specification limits. The smaller the process variation relative to the specification limits, the better SPC will work.

Shewhart proposed a process, known as the PDCA cycle, to drive down process variation so SPC would become more and more effective with any given process. To this day, his variation reduction cycle is known as the Shewhart cycle, or the Deming cycle. The PDCA cycle stands for Plan - Do - Check - Act. To reduce the variation in any process, the analyst must plan, decide what action might reduce process variation; do, try out the idea; check, determine with data that the process variation idea was effective in reducing variation; act, implement the idea permanently (see Chapter 9). Upon conclusion of the cycle, another idea would be tried, and the cycle repeated. This variance reduction process would continue. The repeated application of the PDCA cycle to a process is known as Continuous Quality Improvement.

Deming's contribution to the CQI/TQM philosophy was to expand upon Shewhart's SPC and PDCA ideas and develop a philosophy of management and management practices that would make this idea work in the real world.

The reader must notice at this point, that the focus of CQI/TQM is on individual processes, one by one, not entire systems.

What is BPR All About?

The credit for articulating what Business Process Reengineering is all about should be given to Michael Hammer and James Champy who articulated many ideas and pieces of BPR in a cohesive theme in their book *Reengineering the Corporation* in 1993.

The primary ideas in BPR are two-fold:

- to move from many small processes that span several departments and reporting lines within a business and which for that reason are not under any single management control to much larger, more complex, processes in which the process is under singular management control; e.g., the business is organized by process, rather than by department
- to reanalyze business processes to remove as many steps that don't add value as possible

Clearly, BPR will have a profound impact on a business. The business will have to be managed differently from a departmental perspective to a process perspective. Fewer people will be working in the organization, because steps that didn't add value are now gone. The remaining people will have to be both smarter (because the process is more complex) and will work harder (non-value added steps are now gone).

The gains in productivity that reengineering can effect are truly large. Data indicates that simply reengineering a typical unplanned business process will double the efficiency of the process without applying any new technology. Frequently in reengineered organization, information technology is introduces as part of the reengineered process, sometimes replacing people. The efficiency gains in this situation frequently quadruple.

The reader should notice at this point, that the focus of BPR is on entire systems, not individual processes that make up the larger systems.

8.8 SIX SIGMA QUALITY

In 1988, the Motorola Corporation was the winner of the Malcolm Baldrige National Quality Award. Motorola bases much of its quality effort on what its calls its "6-Sigma" Program. The goal of this program was to reduce the variation in every process to such an extent that a spread of 12ς (6ς on each side of the mean) fits within the process specification limits. Motorola allocates 1.5ς on either side of the process mean for shifting of the mean, leaving 4.5ς between this safety zone and the respective process specification limit.

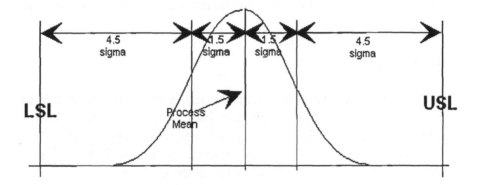

Figure 8.15 Six Sigma Quality.

Thus, as shown in Figure 15, even if the process mean strays as much as 1.5ς from the process center, a full 4.5ς remains. This insures a worst case scenario of 3.4 ppm nonconforming on each side of the distribution (6.8 ppm total) and a best case scenario of 1 nonconforming part per **billion** (ppb) for each side of the distribution (2 ppb total). If the process mean were centered, this would translate into a C_p=2.00.

Companies like GE and Motorola have made significant progress toward this goal across most processes, including many office and business processes as well.

Six Sigma is a rigorous, focused and highly effective implementation of proven quality principles and techniques. Incorporating elements from the work of many quality pioneers, Six Sigma aims for virtually error free business performance. Sigma, s, is a letter in the Greek alphabet used by statisticians to measure the variability in any process. A company's performance is measured by the sigma level of their business processes. Traditionally companies accepted three or four sigma performance levels as the norm, despite the fact that these processes created between 6,200 and 67,000 problems per million opportunities! The Six Sigma standard of 3.4 problems per million opportunities is a response to the increasing expectations of customers and the increased complexity of modern products and processes.

8.9 SUMMARY

Process capability charts may actually be a misnomer. The graphics presented with this feature include a histogram of the data, with a distribution curve overlaid. The statistics provided largely deal with the capability indices. Capability indices attempt to indicate, in a single number, whether a process can consistently meet the requirements imposed on the process by internal or external customers. Much has been written about the dangers of these estimates, and users should interpret capability only after understanding the inherent limitations of the specific index being used.

Process capability attempts to answer the question: can we consistently meet customer requirements? The number one limitation of process capability indices is that they are meaningless if the data is not from a controlled process. The reason is simple: process capability is a prediction, and you can only predict something that is stable. In order to estimate process capability, you must know the location, spread, and shape of the process distribution. These parameters are, by definition, changing in an out of control process. Therefore, only use Process Capability indices if the process is in control for an extended period.

The same argument holds for a histogram. If the purpose of creating a histogram is to see the "shape" of the process, it will be very misleading if the process is not stable. For example, a process that is stable could very well have the same histogram as a process undergoing a trend, since the only difference in the data would be the order of the points. Since the histogram does not consider the sequence of the points, you would see no difference between histograms.

Six Sigma is a quality improvement methodology structured to reduce product or service failure rates to a negligible level (roughly 3.4 failures per million opportunities). To achieve these levels, it encompasses all aspects of a business, including management, service delivery, design, production and customer satisfaction.

REFERENCES

Ale, B. (1991), "Acceptability Criteria for Risk and Plant Siting in the Netherlands," VIII Meeting 3ASI: Safety Reports in EC, Milan, September 18th-19th.

Apostolakis, G., Garrick, B. J. and Okrent, D. (1983), "On Quality, Peer Review, and the Achievement of Consensus in Probabilistic Risk Analysis," Nuclear Safety, Vol. 24, No. 6, November/December, pp. 792-800.

Breyfogle, F. W. (1999), Implementing Six Sigma: Smarter Solutions using Statistical Methods, Wiley & Sons, New York, N.Y.

Chang, S. H., Park, J. Y. and Kim, M. K. (1985), "The Monte-Carlo Method Without Sorting for Uncertainty Propagation Analysis in PRA," Reliability Engineering, Vol. 10, pp. 233-243.

Committee on Public Engineering Policy (1972), Perspectives on Benefit-Risk Decision Making, National Academy of Engineering, Washington DC.

Eckes, G. (2000), General Electric's Six Sigma Revolution: How General Electric and Others Turned Process Into Profits, John Wiley & Sons, New York, N.Y., November.

Gorden, J. E. (1978), "Structures: Or Why Things Don't Fall Down," Da Capo Press, New York, N.Y.

Joksimovich, V. (1985), "PRA: An Evaluation of the State-of-the-Art," Proceedings of the International Topical Meeting on Probabilistic Safety Methods and Applications, EPRI NP-3912-SR, pp. 156.1-156.10.

Kaplan, S. and Garrick, B. J. (1981), "On the Quantitative Definition of Risk," Risk Analysis, Vol. 1, No. 1, pp. 11- 27.

Kyburg, H. E. Jr. and Smokler, H. E. (1964), Studies in Subjective Probability, John Wiley & Sons, New York, NY.

Lichtenberg, J. and MacLean, D. (1992), "Is Good News No News?", The Geneva Papers on Risk and Insurance, Vol. 17, No. 64, July, pp. 362-365.

March, J. G. and Shapira, Z. (1987). "Managerial Perspectives on Risk and Risk Taking," Management Science, Vol. 33, No. 11, November, pp. 1404-1418.

Morgan, M. G. and Henrion, M. (1990), "Uncertainty: Guide to dealing with Uncertainty in Quantitative Risk and Policy Analysis," Cambridge University Press, Cambridge, U.K.

Roush, M. L., Modarres, M., Hunt, R. N., Kreps, and Pearce, R. (1987), Integrated Ap-proach Methodology: A Handbook for Power Plant Assessment, SAND87-7138, Sandia National Laboratory.

Rowe, W. D. (1994), "Understanding Uncertainty," Risk Analysis, Vol. 14, No. 5, pp. 743-750.

Schlager, N., ed. (1994), When Technology Fails: Significant Technological Disasters, Accidents, and Failures of the Twentieth Century, Gale Research, Detroit, M.I.

Shafer, G. and J. Pearl, ed. (1990), Readings in Uncertain Reasoning, Morgan Kaufmann Publishers Inc., San Mateo CA.

Slovic, P. (1987). "Perception of Risk," Science, Vol. 236, pp. 281-285.

Slovic, P. (1993), "Perceived Risk, Trust, and Democracy," Risk Analysis, Vol. 13, No. 6, pp. 675- 682.

Wahlstrom, B. (1994), "Models, Modeling and Modellers: an Application to Risk Analy-sis," European Journal of Operations Research, Vol. 75, No. 3, pp.477-487.

Wang, J. X. and Roush, M. L. (2000), What Every Engineer Should Know About Risk Engineering and Management, Marcel Dekker Inc., New York, NY.

Wenk, E. et. al. (1971), <u>Perspectives on Benefit-Risk Decision Making</u>, The National Academy of Engineering, Washington DC.

Wheeler, D. J., Chambers, D. S. (1992), <u>Understanding Statistical Process Control,</u> SPC Press, Knoxville, T.N., June.

Zadeh, L. A. and Kacprzyk, J., ed. (1992), <u>Fuzzy Logic for the Management of Uncertainty</u>, John Wiley and Sons, New York NY.

9

Engineering Decision Making: A New Paradigm

The nature of business has changed rapidly in the last 25 years from a production-focused orientation to a customer-oriented one. Engineering decision making must recognize this trend. Competition for the customer has increased dramatically with the rapid development of the worldwide engineering products. The pace of technological change has accelerated its effect on today's engineering activities. Concurrently, many new developments in engineering decision-making theory and practice have occurred to address the challenges more effectively.

9.1 ENGINEERING DECISION MAKING: PAST, PRESENT, AND FUTURE

For most of the 19th and 20th century, since the industrial revolution, engineering decision-making philosophy and practice has been an extension of the military organizations of ancient Rome. The style of technical organization was a command and control hierarchical style from the top down. Like a military organization, a chief engineering officer decided what the goals and objectives of the product development were to be, and communicated a set of orders to the "captains" under him, and these men led the field soldiers (who did the fighting) into battle.

The strategy was formulated at the top of the command/control structure, the actual design activities took place by people at the bottom of the structure.

The industrial revolution came about in several parts. First came the power systems (wind, steam, and water) that assisted human power in manufacturing. Second, during the time of the American Civil war in the mid-1800s, came the idea of interchangeable parts. A gun, for example, instead of being hand made as an intact entity, would be made of components that were precisely made that would work together. For example, the trigger assembly for a gun would work in any gun that was made with a similar design. The trigger assembly company would thus be a tier one supplier for gun OEMs.

Finally came the idea of assembly line manufacturing and division of labor. The idea here was that people could be more efficient if they would perform only one simple job when manufacturing something (like a car, for example). This idea was put into practice by Henry Ford for making Model-T cars, and perfected soon after that by people such as Fredrick Taylor and the scientific school of engineering decision-making theory.

Since the military was really the only decision-making model for a large organization, it quite naturally became the organizational model for engineering and manufacturing in the United States. In fact, military commanders were frequently chosen to run large engineering organizations because of their familiarity with this command/control hierarchical structure.

For most of the 20th century, until the mid-1980's, this philosophy of decision-making prevailed in America and this is what was taught in engineering management classes in the United States.

Nineteenth century engineering decision-making philosophy began to unravel during the late 1970s when the world (particularly the United States) found that it had serious competition. Since the late 1970s, the big-three's (Ford, GM, and Chrysler) share of the worldwide automobile market has fallen by half. The United States had 100% of the world television and VCR market at one point (both were invented in the U.S.) and now produces no TVs or VCRs at all. An American car made in America had about $1000 more in overhead (management) than a Japanese car made in Japan. Japan's cars were of much higher quality, had fewer defects, lasted longer, and cost less than comparable American cars. Model changeover in an American car line took 6 weeks while the Japanese did it in 3 days.

Engineering decision making is the primary problem. Since 1990, a single truth that affects engineering decision making has become evident. The manufacturing climate has changed from production driven systems to customer driven systems. Up until the mid-1980s, manufacturers were more driven to achieve production efficiencies than customer service and the focus of the enterprise was production efficiencies and unit cost reduction. Since 1990, with worldwide competition, the situation has changed to one in which the customer's desires drive the enterprise. On-time delivery and product and service quality have become highly important, and the continuing need for innovative product characteristics have fueled a continuing cycle of product improvement and change. This is particularly true in the computer industry.

For example, Toshiba upgraded its entry level Toshiba 100 series notebook computer five times in eleven months while IBM has a strategy of completely revamping its entire notebook product line every six months. Why do companies do this? Because, if they don't, their competitors will get the customer's business. And yes, the customers still expects competitive prices.

It became clearly evident that top-down, hierarchical, multi-layered, production-oriented, manufacturing strategies simply didn't work anymore. Having workers specialize in a single operation and hand off to the next person who specialized in another single operation while effective for cost efficiencies given unlimited production, simply didn't work in an environment in which the customer demanded product attributes other than lowest cost.

Not only did engineering have to change how they worked, engineering decision making had to change as well. A new-generation engineering decision making is the result.

9.2 ENGINEERING DECISION MAKING TOOL BOX

An Engineering decision making tool box includes many product development and manufacturing concepts that have been developed and implemented between the 1930's and the present time. Only recently have they been integrated into a cohesive management philosophy. The components of the new generation engineering decision making includes:

- Statistical Process Control (see Chapter 8)

- Continuous Improvement and Total Quality Management

- Customer Focus

- The Theory of Constraints

- Activity-Based Costing (ABC)

- Just-In-Time Production/Inventory Systems (JIT)

- Strategic Planning (including mission, value, and vision statements)

- Business Process Re-Engineering

- Benchmarking and Best Practices

- External Validation Models (ISO-9000, QS-9000, the Deming Prize, and the Baldrige Award)

• Learning Organization Concepts

Many of these concepts have been elegantly integrated into a cohesive measurement/decision-making theory called the *balanced scorecard* by Harvard Business School professor Robert Kaplan and management consultant David Norton (Kaplan & Norton, 1993) (Kaplan & Norton, 1996). As shown later in the chapter, the balanced scorecard is being applied to engineering decision-making process.

Following is a brief description of each of the constituent elements of new generation engineering decision-making:

Statistical Process Control (SPC)

When products were individually handmade, each part was made to individually work with all other parts *in that individual product*. With the advent of mass production, each part had to work *in all products of its type*. Thus the parts were interchangeable. Quality, particularly the dimensions of component parts, became a very serious issue because no longer were the parts hand-built and individually fitted until the product worked. Now, the mass-produced part had to function properly in every product built. Quality was obtained by inspecting each part and passing only those that met specifications.

Processes and Process Variability

Process variation can be partitioned into two components. Natural process variation frequently called common cause or system variation, is the naturally occurring fluctuation or variation inherent in all processes. In the case of the basketball player in Chapter 8, this variation would fluctuate around the player's long-run percentage of free throws made. Some problem or extraordinary occurrence typically causes special cause variation in the system. In the case of the basketball player, a hand injury might cause the player to miss a larger than usual number of free throws on a particular day.

Continuous Improvement and Total Quality Management

Shewhart developed a never-ending approach toward process improvement called the Shewhart Cycle (also known in Japan as the Deming Cycle and most frequently today in the United States as the Plan-Do-Check-Act or PDCA Cycle). This approach emphasizes the continuing, never-ending nature of process improvement (see Chapter 8.)

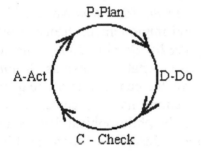

Figure 9.1 A Plan-Do-Check-Act or PDCA Cycle

During the early stages of World War II, America had a substantial problem with military ordnance (bombs, bullets, torpedoes, etc.). A team, called the whiz kids--including Shewhart, was put together as an attempt to improve ordnance performance. The team was highly successful in further refining SPC and utilizing it to improve the quality of weaponry.

After World War II, Deming and Shewhart presented lectures on SPC and the wartime refinements to it to American engineers and managers involved in manufacturing. Many engineers who heard were convinced and adopted the SPC philosophy. Most managers were more interested in meeting the worldwide pent-up demand for products and concentrated on making more product rather than better product. Thus, the management implications were largely ignored in America.

Both Deming and Shewhart were selected to go to Japan and participate in Japanese reconstruction efforts. They presented their lectures in Japan. The Japanese managers and engineers listened and adopted the philosophy. Within five years they had internalized the philosophy. By the early 1970's, the United States had begun to feel the competition. By the early 1980's the quality gap between Japan and the United States had become critical.

In 1982, Deming published his books *Out of the Crisis* and *The New Economics for Industry, Government, Education* which laid out his application of process control and improvement theory to management (Deming 1982a, Deming 1982b, Deming 2000). The concepts in these books are summarized in what is known as *Deming's 14 Points*. The 14 Points have become the centerpiece of the quality philosophies of many companies. Deming has two contemporaries who also have contributed greatly to the philosophical development of the quality improvement movement, Joseph M. Juran and Philip B. Crosby. Although their perspectives and approaches are somewhat different, their philosophies have more similarities than differences and all are rooted in the philosophy of continuous process improvement.

Today, most organizations that have successfully internalized a philosophy of continuous improvement have followed the approach of Deming, Juran, or Crosby. These organizations include manufacturing, service, health, education, military and government agencies. As a result of continuous improvement, America is once again becoming competitive.

TQM/CQI Philosophy and Practices

Today, a TQM-based management philosophy includes not only the core TQM/CQI concepts but also supporting concepts directly related to continuous improvement. These supporting concepts are natural extensions of, supportive of, and in many cases prerequisite to, successful implementation of a TQM/CQI philosophy in an organization.

Central Core Concepts

- The concept of a system and systems analysis
- Process variability, including common cause and special cause variation
- Statistical process control and control charts to identify special cause variation
- PDCA cycle to improve processes continuously by reducing common cause variation
- Tools to identify the root cause problems of processes and to assist in implementing new processes

Supporting Concepts

- Emphasis on customers, both internal and external, and their needs
- Employee issues including:
  - Empowerment
  - Teams, including cross-functional and self-directed teams
  - The value of employees
  - Emphasis on education and training

The Concept of a System

Deming defines a system as the entire organization (Deming 1982a, Deming 1982b, Deming 2000). Deming's system is composed of management; management philosophy; employees; citizens of the country; all facets of government, including laws, taxes, trade barriers, et cetera; foreign governments; customers; shareholders; suppliers; environmental constraints; and banks and other financial entities. He believed strongly that the components of the system should work together like a symphony orchestra. He even believed that competitors formed part of the system and so taught the Japanese in the early 1950s. This is probably a major reason for the synchronization and cooperation of Japanese industry and America's difficulty in penetrating that market today.

Customer Focus

It has become very clear, particularly since the late 1980s that more and more research is directed at customers and this data is used to improve the performance of the company in areas that the customer deems important. This includes several types of analyses:

- Traditional market research to assess the probability of future success of new products and services is being performed more frequently. An interesting offshoot of this is that the market research today has a focus of determining the demographic characteristics of the potential market for a product. And, as a consequence, the product may be targeted to a particular market segment. Or, the product may be modified to appeal to a particular market segment.
- Customer satisfaction analysis is being very frequently applied in most service businesses and for an increasing numbers of manufactured products. Today, not only is the satisfaction for the actual product measured, but also the marketing, sale, and service of the product are measured.

These data are then used to improve both the product and the services surrounding the product to continuously improve the customer's relationship with the company.

Teams

In American manufacturing, what was formerly done with individual high-speed manufacturing processes with an individual working repetitively on some operation on a part has now changed to an automated system where the individual operations are performed by automated machines and robots with many machining steps performed in a machining center. Quality is measured frequently during the manufacturing cycle. Clearly the role of the worker has changed from putting the part into the machine and pressing the "Cycle Start" button to that of a problem-solver, capable of recognizing production processes and solving them.

Most frequently, these teams are self-managed, with responsibility for process performance, both quality and quantity, assigned to the team. The team is also empowered to make decisions related to the process and most frequently has control over budget and personnel. When a process problem arises, the team has well-developed problem solving skills to assist in solving the problem. Frequently they are also trained in Kaizen--Japanese workplace management that is designed to promote workplace efficiency and order.

Companies placing a significant reliance on teams also place a great deal of emphasis on team training. The American automobile maker Saturn invests approximately 140 hours per employee per year to team training. This is in addition to the technical training necessary to function with their process. Clearly, they have decided that the economic gains achieved from team training are worth the cost in lost production time and money.

Another type of team is the cross-functional team. It is employed primarily in the design phase of a new product or process. When a new product is anticipated, a cross-functional team is created to bring the product to market. Cross-functional teams include members from all relevant portions of the enterprise including production, marketing, design, etc. Thus, products are created that can be both marketed and produced effectively. The management of cross-functional teams requires training both for the team leader as well as team members to be truly effective.

The Theory of Constraints (TOC)

Eli Goldratt first formulated *the theory of constraints* approximately 10 years ago with the publication of his novel, *The* Goal (Goldratt and Cox, 1992). *The Goal* was written as a novel to teach the concept of employing the removal of constraints from systems (including production systems) to facilitate achieving the singular most important goal of a business. Subsequent to that time, Goldratt has written several other books that clarify and systematize theory of constraints thinking (see Chapter 5, section 5.5).

Goldratt considers a business to be a financial system, whose goal is to produce more money now and in the future. Goldratt defines *Throughput* as the rate at which the firm produces money. The major idea is to continuously maximize the *Throughput* of the business. This is done by continuously monitoring (with a real-time information system) the flow through the system (either money or production). This allows the removal of bottlenecks to achieving maximum throughput.

Activity-Based Costing (ABC) and Just-In-Time Production/Inventory Systems (JIT)

When production decisions are made using the theory of constraints, the result is called JIT or just-in-time manufacturing. When financial decisions are made with this theory, it is called ABC or activity-based costing. Interestingly enough, the Japanese car manufacturer Toyota discovered JIT more than 15 years ago, but never could explain why it worked. TOC now provides the underlying theory.

TOC clearly has applications in workflow management, continuous improvement in both production and service environments, and in project management. These applications of TOC are just being developed today.

Strategic Planning (including mission, value, and vision statements)

The concept of strategic planning has been a focus of business since the 1970s. Many articles and books have been written on the subject. A concise listing of 1990s thinking in strategic planning and business strategy is found in the three *Harvard Business Review* articles and Hamel and Prahalad book listed in the references.

Today, most businesses have at least one of three documents that define what the business is and its future aspirations.

The *mission statement* states the essence of what the organization is all about. It serves to articulate how this business is different from all other businesses and its competitors. It places limits on the scope of the business. In some cases, it may state the core competencies that the business deems important. It is frequently a rather inspirational document.

The *value statement* articulates the values and ethical principles the company believes in. Frequently one finds statements about the environment, the company culture, employee entrepreneurship, treatment of company personnel, affirmative action, team functioning, or belief in quality as part of these statements. Frequently, top management and company employees jointly produce the value statement.

The *vision statement* sets forth the organization's view of where it will go and what it will look like in the future. It articulates a vision or desired scenario of what the company will look like, its characteristics, its culture, its business practices, and its profitability.

The *strategic plan* is a rolling document that translates the vision into global operational terms frequently complete with long-term goals and measures to actualize the vision within a fixed time frame and a methodology to determine progress. Frequently the strategic plan is a view of 5-7 years into the future.

The *operational plan* is a series of short-term (1-2 years) steps that accomplish steps in the strategic plan. Progress is measured much more frequently (monthly or quarterly) to determine the extent of progress.

The goal of strategic planning is to translate the corporate mission, vision, and value statements into long-term reality.

Business Process Re-Engineering

The term *business process re-Engineering* was coined by Champy and Hammer with their book of the same title in 1993.

As U.S. businesses have moved from a traditional manufacturing environment that is production driven to one that is customer driven, it has become clear that business processes had to drastically change to improve performance in areas that customers deemed important. To do this required businesses to rethink how their processes served customers. It became immediately clear that the production model of business, with a one person doing a small operation on a product, then handing it off to someone else, was not a model that promoted quick cycle times or control of the many individual processes because too many people were involved, none of whom had responsibility for the entire process. In fact, the many handoffs that occurred frequently crossed departmental lines.

Business process reengineering (BPR) is a methodology (using systems analysis) to create processes that have the smallest number of handoffs possible with responsibility (and authority) for the entire process vested in one person or one team. Of necessity, this requires people of a higher skill level than in a production-driven environment. A second goal of BPR is to reduce the number of non-value added activities to as close to zero as possible. Non-value added activities are activities that do not directly impact the production of the product or service. Management is a non-value-added activity as are quality checks, expediting, etc. When BPR is applied, frequently cycle times decrease by 60%-70% and more. Frequently costs are cut by similar amounts. At the start of a successful BPR effort, training must take place as much more sophisticated workers are prerequisite to success.

Benchmarking and Best Practices

Benchmarking is a strategy developed in the early 1980s of comparing one's business to recognized world-class competitors to discover practices and processes used in these businesses that make them successful. In some cases what is discovered is technological advantages, in some cases it is management practices, and in other cases it is process innovations. The goal is to learn from your competitor and use their learning to stimulate similar advances in your business.

A good example is Motorola. In the early days of cellular telephone and pager manufacturing, approximately 30% of Motorola's products were defective when they were about to be shipped or arrived DOA (dead on arrival) at the customer. In Japan and Germany, however, this rate was many times smaller. Why?

Motorola sent a team of management, technicians, and workers to these plants to determine why. Some actually got jobs in these plants. After approximately one year, they got together and pooled what they had found in technology, manufacturing technology, management practices, design for quality, management of company culture, statistical process control, etc. What resulted was a new manufacturing facility that incorporated many of the ideas found, including the seeds for Motorola's own drive toward quality called the six sigma program. When finally in place, it surpassed all of its competitors within a year producing defective product at such a low rate (4.3 defective phones per million produced) that the final product no longer had to be inspected. The U.S. Department of Commerce maintains listings of best practices in various fields that all manufacturing and service companies can access on the Internet.

External Validation Models (ISO-9000, QS-9000, the Deming Prize, and the Baldrige Award)

ISO-9000 is a world wide standard for documenting the processes used in manufacturing or providing services to insure to customers that the manufacturing, quality, and service standards advertised by the company are true. Thus, a purchaser is insured of at least minimal performance with respect to quality and service.

QS-9000 is an American extension of ISO-9000 that includes a criteria for continuous improvement. Typically, an American company would first achieve ISO-9000, then work toward QS-9000.

The Baldrige is America's highest award for quality and is given to two companies (one large and one small) each year (out of many hundreds that apply) that compete for the award. The award is named for former Secretary of Commerce Malcomb Baldrige. The award is given by the U.S. Department of Commerce and presented by the President of the United States. The award is based on a complex criteria of management, quality, culture, and customer service criteria. Most companies who apply realize they will never win, but use the effort to focus their drive to be a high performance company because the measurement criteria truly provides an index of how good the company is on a large variety of important criteria.

The Deming Prize is Japan's highest award for quality. It is named for American statistician W. Edwards Deming who brought the concepts of quality, SPC, and continuous improvement to Japan after World War II. Many of the criteria is similar to that of the Baldrige award, but it is oriented to companies who actively use statistical process control charting in their quality efforts. Only one U.S. company has ever won the Deming Prize, Florida Power and Light.

The major philosophical difference between the Baldrige Award and the Deming Prize is that the Deming Prize is awarded to all companies who meet the criteria while the Baldrige Award is awarded only to the best large and the best small companies each year.

These awards are important validation that a company's efforts to become world-class in quality are making progress. They identify weak areas in the company so they can be strengthened. It's not the winning of the award that's important. It's the journey toward the award, getting better and better in better in every way.

Learning Organization Concepts

Peter Senge coined the term "learning organization" in his book *The Fifth Discipline* in 1990. A learning organization is an organization that recognizes the importance of continuous learning by employees so the organization can be responsive to the rapidly changing marketplace that exists in a global economy. The world today is information driven, and the successful organization of the future will consist of an intelligent, well-educated workforce that can use information to make business decisions, that effectively gathers and uses information in its business, and that develops effective strategies for harnessing the mind power of each employee in the organization.

The goal today is to solve problems close to the source and to harness the brainpower of each person. The thinking is that the organization should function as a large parallel-processing computer, harnessing the minds of each employee. Senge argues that when this is well done, the consequences will include a much more powerful organization that is much more capable of change and of taking advantage of opportunities. He sees well-trained teams as a primary mechanism for harnessing this collective intellectual resource.

9.3 BALANCING TECHNICAL MERITS, ECONOMY, AND DELIVERY

The important question really is how to integrate all of these new ideas into a cohesive measurement and management strategy to ensure that progress is being made in each important area. In fact, it is necessary to define what the important areas are. Professor Robert Kaplan of the Harvard Business School and well-known management consultant David Norton have, based on extensive research among American businesses, have identified four major perspectives that should be measured and managed in business today. As described briefly in the previous section, these concepts have been elegantly integrated into a cohesive measurement/management theory called the *balanced scorecard.*

The four perspectives chosen by Kaplan and Norton are:

1. Financial
2. Customer
3. Internal Business Processes

4. Learning and Growth

Kaplan and Norton define two types of measures: *driver measures* and *outcome measures*. Outcome measures are the more traditional type of business financial measure, such as return on investment (ROI). It is useful in historically documenting business performance over time, but it is useless as a management tool, by itself, to pinpoint the location of a problem area within a business. To do this takes a driver measure. One such measure might be a customer satisfaction measure.

Typically outcome measures are termed *lagging* measures while driver measures are called *leading* measures. Typically, if a serious problem is detected in a leading measure, it will also manifest itself later in some outcome measure that lags behind.

The goal of this measurement/management philosophy should now become clear. It is to measure the appropriate business variables with *leading* measures so that specific remedial action can be taken early on, long before the problem would appear in a traditional *lagging* measure such as ROI.

Kaplan and Norton suggest that a series of cause-effect relationships exist through the processes of a company between appropriately selected leading measures and the lagging company financial performance measures. The nature of these cause-effect linkages operates something like the following example:

Example 9.1: Improving process capability is prerequisite to both process quality and process cycle times. These are both essential to on-time delivery of a high quality product, both of major importance to customer loyalty. Customer loyalty is prerequisite to having a satisfactory Return on Capital Employed (ROCE). This can be illustrated by the diagram in Figure 9.2.

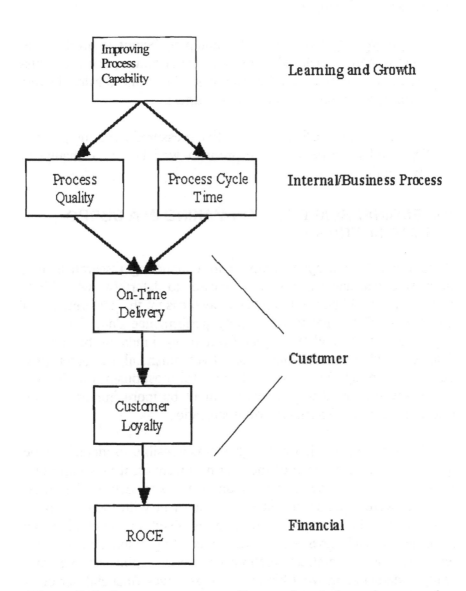

Figure 9.2 Cause-consequence diagram for a balanced scorecard.

Here is another example.

> *Example 9.2:* A process in the company begins to produce defective parts which, when used by the customer, cause the customer to buy from another supplier. This reduces sales, lowers profits, and reduces return on investment.

Clearly, using a driver measure that measured outgoing product quality could have detected the problem well before a change in ROI was noticed.

9.4 ENGINEERING DECISION MAKING IN A CORPORATE SETTING

Kaplan and Norton spent many years conducting research among both American and worldwide businesses to determine what kinds of variables could be used to adequately measure the performance of companies. They analyzed company performance data. They surveyed executives and managers from many kinds of businesses. They concluded that the four perspectives: financial, customer, process, and learning/growth were adequate for most businesses. This is not to say that another perspective might be more important than one of these in a different type of organization.

They concluded that although it was possible to identify some generic measures on each of the four perspectives, it was important for each business to perform a custom analysis to identify the measures that were most appropriate and congruent with their company mission and direction. The learning involved in developing the measures typically goes a long way in assisting management in determining the cause-effect relationships between the variables and really understanding what kinds of things drives financial performance. They also concluded that a balance between leading and lagging indicators was important.

Table 9.1   Financial Measurements for a Balanced Scorecard

Business Maturity	Revenue Growth and Mix	Cost Reduction/ Productivity Improvement	Asset Utilization
Growth Phase	Sales growth rate by segment Percentage revenue from new product, service, and customers	Revenue/ employee	Investment (% of sales) R&D (% of sales)
Sustain Phase	Share of targeted customers and accounts Cross-selling Percentage revenues from new applications Customer and product line profitability	Cost versus competitors' Cost reduction rates Indirect expenses (as % of sales)	Working capital ratios ROCE by key asset categories Asset utilization rates
Harvest Phase	Customer and product line profitability Percentage unprofitable customers	Unit costs (per unit of output, per transaction, etc.)	Payback Throughput

Financial

Kaplan and Norton believe (based on supporting research) that the important financial themes for a company depend to a great extent on the maturity of a business. When a business is just beginning, different measures should be employed than when the business is in stable state. Finally, other measures should be employed when the business is in "harvest" stage--that stage when the investments in the prior two stages should be exploited. As shown in Table 9.1, Kaplan and Norton suggest generic measures in three areas:

1. Revenue Growth and Mix

2. Cost Reduction/Productivity Improvement

3. Asset Utilization

Customer

The business exists to serve the customer. Thus, it is critically important that the business understands the customer, the customer's wants and needs, and how well the customer feels the business is serving him. Kaplan and Norton suggest the core measures as shown in Table 9.2.

During the late 1980s through the early 1990s, executives believed that reducing costs was the answer to increased global competition. Businesses downsized, de-layered, re-engineered and restructured their organizations. While this was probably necessary to cut the "fat" out of organizations, it wasn't a formula for long-term success. Today the focus is on delivering more value and increasing customer loyalty.

Customer relationship management (CRM) is a business strategy to select and manage customers to optimize long-term value. CRM requires a customer-centric business philosophy and culture to support effective marketing, sales, and service processes.

CRM applications can enable effective Customer Relationship Management, provided that an enterprise has the right leadership, strategy, and culture.

Table 9.2   Core Measures from Customers' Perspectives

Market Share	Reflects the proportion of business in a given market (in terms of number of customers, dollars spent, or unit volume sold) that a business unit sells.
Customer Acquisition	Measures, in absolute or relative terms, the rate at which a business unit attracts or wins new customers or business.
Customer Retention	Tracks, in absolute or relative terms, the rate at which a business unit retains or maintains ongoing relationships with its customers.
Customer Satisfaction	Assesses the satisfaction level of customers along specific performance criteria.
Customer Profitability	Measures the net profit of a customer, or a market segment, after allowing for the unique expenses required to support the customer.

Internal Business Processes

Kaplan and Norton suggest three critical business processes which take place from the moment a customer need is identified to the time at which that customer need is met. Each of these processes should be measured.

Kaplan and Norton's measures for Internal Business Processes is shown in Table 9.3.

Table 9.3 Measures for Internal Business Processes

Internal Process	Process Steps	Suggested Measures
The Innovation Process	Identify the Market Create the Product or Service Offering	Percentage of sales from new products Percentage of sales from proprietary products New product introductions vs. both competitors and plans Manufacturing process capabilities Time to develop next generation of products
The Operations Process	Build the Product or Service Deliver the Product or Service	Cycle times for each process Quality measures Process parts-per-million defect rates Yields Waste Scrap Rework Returns % of processes under statistical control
The Post-Sale Service Process	Service the Customer	Customer satisfaction surveys % of customers requiring service

Learning and Growth

Kaplan and Norton suggest that that there are three basic components that measure the ability of a company to learn, grow, and keep pace with intellectual competition. They are the competencies of the staff, the sophistication of the technology infrastructure, and the company climate. This can be summarized in the matrix shown by Table 9.4.

Table 9.4 Measures for Learning and Growth

Staff Competency	Technology Infrastructure	Climate for Action
Strategic Skills	Strategic technologies	Key decision cycle
Training Levels	Strategic databases	Strategic focus
Skill Leverage (how well are they used and deployed)	Experience capturing	Staff empowerment
	Proprietary software	Personal alignment
	Patents and copyrights	Morale
		Teaming

9.5 SUMMARY

The nature of engineering has changed rapidly in the last 25 years from a production-focused orientation to a customer-oriented one. Concurrently, many new developments in management theory and practice have occurred. Many, such as TQM or strategic planning claim to be the total answer. If only life was that simple; unfortunately it is not. The solution lies in identifying and measuring a broad spectrum of pivotal business activities and embracing a wide variety of effective new practices and systems; then molding them into strategies that are effective in today's business environment.

Thus, both how we measure business performance and how we manage businesses in the future can no longer be strictly financial lagging measures that are primarily reactive. They must be dynamic measures that allow managers to deal with problems at an early stage before they have significant negative consequences on the financial health of the business.

The balanced scorecard is such a system. As shown in Figure 9.3, Customers are the ultimate judges about the quality of our decisions!

Figure 9.3 Customers are the ultimate judges about the quality of our decisions!

REFERENCES

Brown, M. (1991), <u>Baldrige Award Winning Quality</u>, Quality Resources, White Plains, NY.

Camp, R. C. (1989), <u>Benchmarking: The Search for Industry Best Practices That Lead to Superior Performance,</u> American Society for Quality Control, Milwaukee, WI.

Champy, J. and Hammer, M. (1993), <u>Reengineering the Corporation: a Manifesto for Business Revolution,</u>   HarperBusiness, New York, NY.

Cokins, G, Stratton, A., and Helbling, J. (1993), <u>An ABC Manager's Primer,</u> Institute of Management Accountants, Montvale, NJ.

Collis, D. J. and Montgomery, C. A. (1990), "Competing on Resources: Strategy in the 1990s," <u>Harvard Business Review</u> (July-August), pages 118-128.

Cooper, R. and Kaplan (1991), R. S., "Profit Priorities from Activity-Based Costing," Harvard Business Review (May-June), pages 130-135.

Deming, W. E. (1982a), Out of the Crisis, MIT Center for Advanced Engineering Study, Cambridge, MA.

Deming, W. E. (1982b), The New Economics for Industry, Government, MIT Press, Cambridge, MA.

Deming, W. E. (2000), The New Economics for Industry, Government, Second Edition, MIT Press, Cambridge, MA.

Garvin, D. (1993), "Building a Learning Organization," Harvard Business Review (July-August): 78-91.

Goldratt, E. (1990) The Theory of Constraints, North River Press, Great Barrington, MA.

Goldratt, E. and Cox, J. (1992), The Goal, North River Press, Great Barrington, MA.

Hamel, G. and Prahalad, C. K. (1994), Competing for the Future: Breakthrough Strategies for Seizing Control of Your Industry and Creating the Markets of Tomorrow, Harvard Business School Press, Cambridge, MA.

Heskett, J., Sasser, E., and Hart, C. (1990)., Service Breakthroughs: Changing the Rules of the Game, Free Press, New York, NY.

Heskett, J., Jones, T., Loveman, G., Sasser, E., and Schlesinger, L. (1994), "Putting the Service Profit Chain to Work, " Harvard Business Review (March-April), pages 164-174.

Jones, T. O. and Sasser, E. (1995), "Why Satisfied Customers Defect," Harvard Business Review (November-December), pages 88-89.

Juran, J. M. (1993), "Made in the U.S.A.: A Renaissance in Quality," Harvard Business Review (July-August).

Kaplan, R. S. (1984), "Yesterday's Accounting Undermines Production," Harvard Business Review (July-August), pages 95-101.

Kaplan, R. and Norton, D. (1992), "The Balanced Scorecard--Measures that Drive Performance," Harvard Business Review (January-February).

Kaplan, R. and Norton, D. (1993), "Putting the Balanced Scorecard to Work," Harvard Business Review (September-October).

Kaplan, R. and Norton, D. (1996), "Using the Balanced Scorecard as a Strategic Management System," Harvard Business Review (January-February).

Kaplan, R. and Norton, D. (1996), The Balanced Scorecard: Translating Strategy into Action, Harvard Business School Press, Cambridge, MA.

Katzenbach, J. R. and Smith (1993), D. K., The Wisdom of Teams: Creating the High Performance Organization, Harvard Business School Press, Boston, MA.

Lorsch, J. W., (1995) "Empowering the Board," Harvard Business Review (January-February): 107, pages 115-116.

Mahoney, F. X. and Thor, C. A. (1994), The TQM Trilogy: Using ISO 9000, the Deming Prize, and the Baldrige Award to Establish a System for Total Quality Management, American Management Association, New York, NY.

McNair, C. J., Mosconi, W. and Norris, T. (1988), Meeting the Technological Challenge: Cost Accounting in a JIT Environment, Institute of Management Accountants, Montvale, N.J.

Prahalad, C. K. and Hamel, G. (1990) "Core Competencies of the Corporation," Harvard Business Review (May-June): 79-91.

Rothery, B. (1991) ISO 9000, Grower Publishing Co., Brookfield, Vermont.

Schneiderman, A. (1988), "Setting Quality Goals," Quality Progress (April), 51-57.

Senge, P. (1990), The Fifth Discipline: The Art and Practice of the Learning Organization, Currency Doubleday, New York, NY.

Shewhart, W. A. (1980), "Economic Control of Quality of Manufactures Product/50th Anniversary Commemorative Issue/No H 0509", American Society for Quality; Reissue edition, Milwaukee, WI.

Srikanth, M. and Robertson, S. (1995), Measurements for Effective Decision Making, Spectrum Publishing Co., Wallingford, CT.

Appendix A

Engineering Decision-Making Software Evaluation Checklist

EASE OF USE

- Masks fields to prevent incorrect data
- Context sensitive help with related topics for browsing
- Clear, intuitive forms
- Tutorials
- Documentation
- Guided tour of features
- Control of Gantt Chart format
- Control of screen colors

PROJECT PLANNING

- Resources visible on same sheet as Gantt Chart
- CPM/PERT chart created automatically
- Built-in project hierarchy for work breakdown
- Gantt chart shows % completion, actuals through, etc.
- Variety of time-scale selections and options
- Global, project, and individual resource schedules

- Resource usage independent of task duration
- Many resources can be assigned to a single task
- Resources can be assigned to multiple tasks concurrently
- Ability to use long task names
- Task duration calculated by system to shorten project and maximize resource utilization
- Split tasks (start, stop, start) under user control
- Critical path calculation based on original or revised dates
- Dependency definition diagram
- Ability to reconfigure CPM network manually
- Copy task(s) in project hierarchy
- Easily move or reorder tasks
- Shift task and see effect on resource usage immediately
- Lengthen/shorten tasks interactively
- Automatically create schedules with resource constraints

ENGINEERING RESOURCE MANAGEMENT

- Automatically level resources according to task priorities
- Set resource usage to maximum percentage of availability
- Insert a project or part of a project into another project
- Go-to feature to move around in large projects
- Manage many types of resources
- Fractional hours permitted
- Resource availability histogram
- Discontinuous tasks permitted
- Categorize resources into groups
- Assign resources with a variety of availability options
- Assign resources with a variety of loading patterns
- Automatically assign resources non-uniformly to task to absorb unused resources

- Summarize resources across multiple projects
- Display resource utilization, unused availability, or total availability on screen
- Analyze resource costs by period

PROJECT TRACKING

- Track against the original baseline plan
- Remove or reset baseline
- Track actual/estimated start and end dates
- Track percentage complete
- Track actual use for all types of resources by period
- Use actuals to calculate project variances
- Actual resource usage independent of task duration
- Actual captured even if in excess of availability
- Replan resource usage
- Perform variable loading per resource per task
- Display and compare original and revised plans simultaneously

ANALYSIS AND REPORTING

- Library of standard reports
- Customizable report structure
- Ad hoc query support
- New report design feature
- Spreadsheet format report structure availability
- Earned value and variance reporting
- Ability to sort and select data for reports
- Variance to baseline reports
- Display task status on CPM network

- Report on-schedule vs. late tasks
- Annotate reports
- Preview reports on-line
- Analyze cash-flow projections
- Plot PERT/CPM network
- Export report to other formats including ASCII

MULTIPLE PROJECTS

- User defined levels of detail
- Multiproject resource spreadsheet
- Level resources across multiple projects
- Create a master project from subprojects or parts of subprojects
- Summarize resources across projects
- Establish interproject dependencies
- LAN support for multiple users with file locking

INTERFACES

- Run on standard Windows 3.11, Windows 95, or Windows NT platforms
- Variety of data export options DIF, PRN, dBase, etc.
- Filter export data
- Import/Export to other vendors software

LIMITS

- Large number of maximum tasks
- Large amount of resources/task [$ and number]
- Long project length
- Large number of dependencies

TECHNICAL SUPPORT

- Vendor support program
- Hot line for technical questions
- Clearly stated upgrade policy
- User's group
- Training and consulting available

Appendix B

Four Primary Continuous Distributions

Four primary distributions are used to estimate the variability of project activities. They are normal, uniform, triangular, and exponential (or negative exponential).

Graphically, they look like the following:

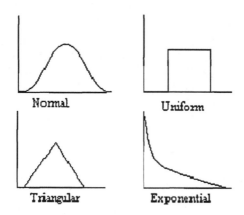

Normal

Uniform

Triangular

Exponential

These random variates are computed in the following manner in Excel:

Normal:

=SQRT(-2*LN((RAND())*2*COS(2*PI*RAND())*STDEV+AVERAGE

where STDEV and AVERAGE are the desired parameters of this set of random variables.

Uniform:

=LOWER+RAND()*(UPPER-LOWER)

where LOWER is the smallest random variate desired and UPPER is the largest.

Triangular:

=IF(RAND()<MO-LO)/(HI-LO),LO+SQRT((MO-LO)*(HI-LO)*RAND())),HI-SQRT((HI-MO)*(HI-LO)*(1-RAND())))

where HI is the rightmost point of the triangle, LO is the leftmost point, and MO is the modal or topmost point. Notice that RAND() **must** be the same random number, so it is computed and then placed into the equation. If the formula is keyed in exactly as above, three random numbers will be used, and the random variate produced will be in error.

Exponential:

The exponential distribution is frequently used in projects to model waiting time for materials or people. Waiting times tend to be heavily positively skew.

=-AVG*LN(RAND())

where AVG is the desired average (a constant, such as 36, for example) for the exponentially distributed random variate and LN is an Excel function to find the natural logarithm.

For more information about the Excel spreadsheet templates and add-in for engineering decision-making analysis, please contact John X. Wang, e-mail: johnjxwang@aol.com

Index

About the Author

JOHN X. WANG is a Six Sigma Quality Master Black Belt certified by Visteon Corporation, Dearborn, Michigan. He is also a Six Sigma Quality Black Belt certified by the General Electric Company. The coauthor of *What Every Engineer Should Know About Risk Engineering and Management* (Marcel Dekker, Inc.) and author or coauthor of numerous professional papers on fault diagnosis, reliability engineering, and other topics, Dr. Wang is a Certified Reliability Engineer under the American Society for Quality, and a member of the Institute of Electrical and Electronics Engineers and the American Society for Mechanical Engineers. He received the B.A. (1985) and M.S. (1987) degrees from Tsinghua University, Beijing, China, and the Ph.D. degree (1995) from the University of Maryland, College Park.